AF894257

Towards New Perspectives on *Toxoplasma gondii*

Edited by Saeed El-Ashram, Guillermo Tellez-Isaias, Firas Alali and Abdulaziz Alouff

Published in London, United Kingdom

Towards New Perspectives on *Toxoplasma gondii*
http://dx.doi.org/10.5772/intechopen.100900
Edited by Saeed El-Ashram, Guillermo Tellez-Isaias, Firas Alali and Abdulaziz Alouff

Contributors
Ruth Adekunle, Almeera Lateef, Helene Santos S. Barbosa, Renata M. De Muno, Marcos A. Moura, Letícia C. Medeiros, Pedro N. Caldas, Rafael M. de Muno, Marialice Da Fonseca da F. Ferreira-da-Silva, Mauricio De Magalhães de Paiva, Erick Vaz Guimarães, María Vanessa Schneider, Gungor Cagdas Dincel, Saeed El-Ashram, Abdulaziz Alouffi, Ebtesam M. M. Al Olayan, Guillermo Tellez-Isaias, Mohammed Hamdy Farouk, Hany M.R. Abdel-Latif, Shagufta Iqbal, Billy M. Hargis, Víctor Manuel Manuel Petrone-García, Gabriela Aguilar-Tipacamú, Inkar Castellanos-Huerta

© The Editor(s) and the Author(s) 2023

The rights of the editor(s) and the author(s) have been asserted in accordance with the Copyright, Designs and Patents Act 1988. All rights to the book as a whole are reserved by INTECHOPEN LIMITED. The book as a whole (compilation) cannot be reproduced, distributed or used for commercial or non-commercial purposes without INTECHOPEN LIMITED's written permission. Enquiries concerning the use of the book should be directed to INTECHOPEN LIMITED rights and permissions department (permissions@intechopen.com).

Violations are liable to prosecution under the governing Copyright Law.

Individual chapters of this publication are distributed under the terms of the Creative Commons Attribution 3.0 Unported License which permits commercial use, distribution and reproduction of the individual chapters, provided the original author(s) and source publication are appropriately acknowledged. If so indicated, certain images may not be included under the Creative Commons license. In such cases users will need to obtain permission from the license holder to reproduce the material. More details and guidelines concerning content reuse and adaptation can be found at http://www.intechopen.com/copyright-policy.html.

Notice

Statements and opinions expressed in the chapters are these of the individual contributors and not necessarily those of the editors or publisher. No responsibility is accepted for the accuracy of information contained in the published chapters. The publisher assumes no responsibility for any damage or injury to persons or property arising out of the use of any materials, instructions, methods or ideas contained in the book.

First published in London, United Kingdom, 2023 by IntechOpen
IntechOpen is the global imprint of INTECHOPEN LIMITED, registered in England and Wales, registration number: 11086078, 5 Princes Gate Court, London, SW7 2QJ, United Kingdom

British Library Cataloguing-in-Publication Data
A catalogue record for this book is available from the British Library

Additional hard and PDF copies can be obtained from orders@intechopen.com

Towards New Perspectives on *Toxoplasma gondii*
Edited by Saeed El-Ashram, Guillermo Tellez-Isaias, Firas Alali and Abdulaziz Alouff
p. cm.
Print ISBN 978-1-80356-494-4
Online ISBN 978-1-80356-495-1
eBook (PDF) ISBN 978-1-80356-496-8

We are IntechOpen, the world's leading publisher of Open Access books

Built by scientists, for scientists

6,300+
Open access books available

171,000+
International authors and editors

190M+
Downloads

156
Countries delivered to

Our authors are among the
Top 1%
most cited scientists

12.2%
Contributors from top 500 universities

CLARIVATE ANALYTICS
BOOK CITATION INDEX
INDEXED

WEB OF SCIENCE™

Selection of our books indexed in the Book Citation Index
in Web of Science™ Core Collection (BKCI)

Interested in publishing with us?
Contact book.department@intechopen.com

Numbers displayed above are based on latest data collected.
For more information visit www.intechopen.com

Meet the editors

Dr. Saeed El-Ashram is a professor at Kafrelsheikh University, Egypt and at Foshan University, China, and a research professor at Zhaoqing Dahuanong Biology Medicine Co., Ltd., China. The primary focus of his research is to understand how the animal immune system recognizes and responds to parasitic infections with and/or without a microbial community. Some of these infections cause significant diseases in humans, such as toxoplasmosis, cryptosporidiosis, alveolar echinococcosis, and fascioliasis, while others, including cryptosporidiosis and coccidiosis, represent a substantial financial burden for food producers. Dr. El-Ashram has over 120 SCI publications, holds several registered patents, and is an academic editor and reviewer.

Dr. Abdulaziz Alouffi is an associate research professor at KACST in Saudi Arabia. His Ph.D. from the University of Nottingham focused on understanding the role of Immunoglobulin E (IgE) in human resistance to infection with metazoan parasites (e.g., *Schistosoma spp*.,) and how this knowledge can be exploited for vaccination purposes. He is currently working on diagnostic methods to identify zoonotic diseases and interventions to mitigate or eradicate them. He is also working with scientists from different countries to broaden the understanding of zoonotic pathogen epidemiology, and how this can be used to inform policy on optimal and cost-effective methods of disease control, and to find new targets for the development of more efficient vaccines against the zoonotic disease.

Dr. Guillermo Tellez-Isaias is a research professor at the Center of Excellence in Poultry Science of the University of Arkansas, USA. He received his DVM and MS degrees in veterinary sciences from the National Autonomous University of Mexico (UNAM), and his Ph.D. from Texas A&M University. He was a professor at UNAM for 16 years, including eight as head of the Avian Medicine Department at the College of Veterinary Medicine. His research focuses on gastrointestinal models in poultry to evaluate the beneficial effects of functional foods on intestinal health and disease resistance. He is a member of the Mexican Veterinary Academy and the Mexican National Research System, and a past president of the National Poultry Science Association of Mexico.

Dr. Firas Alali gained his master's degree from the Department of Veterinary Parasitology at Baghdad University, Iraq, and his Ph.D. from the Faculty of Veterinary Medicine at Selcuk University, Turkey. He has worked in the Department of Veterinary Parasitology at Kerbala University, Iraq, for 10 years, including a year as a vice dean at the College of Veterinary Medicine. He researches in the fields of parasitology, entomology and vector-borne diseases, including zoonoses, with a focus on protozoology, ectoparasites, and the effect of zoonoses on different hosts. He has been a member of the Iraqi Veterinary Syndicate since 2004.

Contents

Preface

Toxoplasma gondii is one of the most common protozoan parasites, affecting 10–25% of the world's population. It has long been thought that toxoplasmosis is a disease mostly affecting neonates who have been infected with the parasite *in utero*. Toxoplasmosis acquired after birth is often an asymptomatic illness that lasts a lifetime. However, in immunocompromised individuals, such as those with acquired immune deficiency syndrome (AIDS) or those receiving organ transplants or intensive cancer treatment, chronic *T. gondii* infections may reactivate, and the result may be fatal. The capacity of *T. gondii* to invade every nucleated cell in humans and the majority of warm-blooded animals is what makes this infection so successful. *T. gondii* grows and lives within the cell after infection. A crucial requirement for lifetime persistence is the transition from the rapidly dividing tachyzoite stage to the quiescent bradyzoite stage.

Towards New Perspectives on Toxoplasma gondii brings together experts from across the world and offers detailed reviews and previously unpublished findings. Topics covered are the development of the schizont stages of *T. gondii* in primary cell culture of feline enterocytes; the effects of toxoplasmosis on host behavior, personality, and cognition; the three nutrient uptake portals in *Toxoplasma* tachyzoites; neuro-immunopathology; and how *Toxoplasma* infection can be diagnosed and prevented.

The editors would like to acknowledge the dedication and hard work of all the authors who have contributed to this book, and the support of the IntechOpen editorial team in bringing this project to completion.

Saeed El-Ashram
Faculty of Science,
Zoology Department,
Kafrelsheikh University,
Kafr El-Sheikh, Egypt

College of Life Science and Engineering,
Foshan University,
Foshan, Guangdong Province, China

Abdulaziz Alouffi
King Abdulaziz City for Science and Technology,
Riyadh, Saudi Arabia

Guillermo Tellez-Isaias
Department of Poultry Science,
University of Arkansas Agricultural Experiment Station,
Fayetteville, AR, USA

Firas Alali
Department of Veterinary Parasitology,
College of Veterinary Medicine,
University of Kerbala,
Kerbala, Iraq

Chapter 1

Introductory Chapter: The Significance of *Toxoplasma gondii* in Humans, and in Domestic and Wild Animals

Saeed El-Ashram, Gungor Cagdas Dincel, Mohammed Hamdy Farouk, Abdulaziz Alouffi, Hany M.R. Abdel-Latif, Shagufta Iqbal, Guillermo Tellez-Isaias, Billy M. Hargis, Víctor Manuel Petrone-García, Gabriela Aguilar-Tipacamú and Inkar Castellanos-Huerta

1. Introduction

1.1 *Toxoplasma gondii* and life cycle

Since 1908, when *Toxoplasma gondii* was isolated in a rodent (*Ctenodactylus gundi*), from which the parasite gained its name [1], more than a century has passed. Because of the crescent-shaped form of the parasite, the genus name was taken from the Greek word toxin, which means bow. It was not until 1970 that the entire life cycle of the parasite was revealed, considering cat (*Felis catus*) the only definitive host (DH) in which the parasite could complete its entire sexual life cycle [2]. *T. gondii* belongs to the *phylum Ampicomplexa*, which contains intercellular parasites with a polarized cell structure and an apical complex [3]. *T. gondii* life cycle is considered sexual and asexual with a characteristic complexity; it reproduces sexually in the DH and asexually in the intermediate hosts (IH) [4] with three distinct stages of reproduction: tachyzoites (replicating fast), bradyzoites (replicating slowly), and sporozoites (in sporulated oocysts). As described briefly, after consuming any of the three stages, wild and domestic cats are infected [5]; prepatent period typically lasts between 3 and 10 days until 19 days or more [6]. Digestive enzymes break down the cyst wall, releasing the bradyzoites [7] and allowing the infection of intestinal epithelial cells through asexual reproduction (schizonts). Merozoites are discharged and reinvade the intestinal epithelium when the schizont reaches maturity. After this stage, sexual differentiation is observed [5, 8], rising to flagellated microgametes and macrogametes and beginning fertilization [6], resulting in the production of a zygote (diploid) and growing into an unsporulated oocyst, which is discharged into the intestinal lumen and expelled from the DH. Sporulation of the oocysts takes from 1 to 5 days, depending on environmental variables such as temperature and humidity [9, 10]. Oocysts

 IntechOpen

have impermeable walls and are resistant to physical and chemical stimuli, living in damp soil and staying infectious for months [11]. The asexual cycle of *T. gondii* begins when an IH consumes contaminated water or food. In the gut, the sporozoites are released and infect the intestinal epithelial cells [5]. The IH may also gain *Toxoplasma* infection by consuming raw or undercooked meat, harboring bradyzoite-bearing tissue cysts. The freshly released bradyzoites convert into tachyzoites after entering the intestinal epithelium after the rupture of the cyst wall [12]. After successful cell invasion, intracellular proliferation takes place until the host cell can no longer sustain its development [13]. Eventually, the tachyzoites egress from the host cell and disperse through the circulatory system to organs such as the liver, lungs, brain, eye, placenta, and heart, completing the asexual life cycle [14]. Once the host's immunological response is elicited, the multiplying tachyzoites develop into bradyzoites, form tissue cysts, and stay dormant in the host [15, 16]. In healthy individuals, tissue cysts may survive throughout life without eliciting an immune response from the host. However, when the immune system is compromised or weakened, bradyzoites may revert quickly to replicating and invasive tachyzoites, leading to infection [13].

1.2 *Toxoplasma* genotypes

T. gondii genome is composed of a 65 Mb nuclear genome with 14 chromosomes [17]. The genome of *T. gondii* is bigger than that of other *apicomplexans* [18], which might be attributed to the large number of IHs of *T. gondii* [19]. Online database (ToxoDB (http://toxodb.org/toxo/)) contains complete information on the genome of *T. gondii* [20], including the three main types of *T. gondii*: GT-1 (type I), Me49 (type II), and VEG (type III) [21, 22]. Genotypes vary substantially across different geographic locations; Asia, Europe, and North and South Americas present type I, and all the continents present types II and III [23]. Thus, diversity in pathogenicity and virulence is directly correlated with genetic differences [24]. A median lethal dose (LD_{50}) to mice in laboratory settings is a common way to describe virulence strain. LD_{50}, for as little as one parasite, in type I strain of the parasite may kill mice. Types II and III, in contrast, are less virulent and only result in LD_{50} following infection with several parasites. Multi-locus restriction fragment length polymorphism (RFLP) is used to detect strains with distinct genotypes compared to types I, II, and III, which were first found in isolated populations of exotic animals, but they have lately been reported in chickens, pigs, and sheep [25]. Human primary infections of ocular illness are associated with atypical toxoplasmosis strains [26]. These atypical strains are characterized primarily by unique alleles and allele-shuffled combinations from the three major types [27]. Microsatellite genotyping of immunocompromised patients with toxoplasmosis revealed three novel genotypes, Africa I, Africa II, and Caribbean I, and subsequently determined a different Africa III genotype [28, 29]. The impact of secreted *Toxoplasma* virulence effectors and genetic diversity of *T. gondii* vary in encystment rates and acute pathogenicity in mice [30]. *T. gondii* strains grouped lineages (types I, II, and III) are based on variations in acute virulence in mice and genetic markers, RFLPs, and single-nucleotide polymorphisms (SNPs) [31, 32]. Highly pathogenic type I strain presents a LD_{100} of 1 [31], in contrast to strain type II with a LD_{50} of 10^3–10^4, and type III strain with a LD_{50} of 10^4–10^5 [33]. Advances in sequencing capabilities resulted in a broader diversity of *T. gondii* strains, giving rise to a new classification scheme, which separates strains into 15 genetic haplotypes, with types I, II, and III being referred to as haplotype 1, 2, and 3 and clade A, D, and C, respectively [33]. Strain-specific alleles of two proteins, ROP5 and ROP18, were found to explain the differential in virulence [34, 35]. The virulent alleles of these proteins operate together to inhibit families of interferon-inducible gene families [35, 36].

1.3 Cats' susceptibility to *Toxoplasma* infection: determinants

Even though DH presents more than 180 parasite species [37], *T. gondii* appears more often than others in literature because of its effects on the health of humans, livestock, and wildlife [38]. Interactions between the three essential "ecological components," (a) DH, (b) environment (Env), and (c) IH influence the sustainability of these parasites, resulting in the fecal-oral route as the most common mode of transmission. A rise in parasite shedding and environmental contamination should be positively associated with the number of DHs. The likelihood of oocysts contributing to parasite transmission may also be influenced by specific host behaviors, and by physically separating the DH and IH niches, it may be possible to reduce parasite transmission. Env conditions may impact *Toxoplasma* oocyst survival, while the behavior of IH (such as grazing) may affect transmission [39]. Oocysts have the potential to survive in the environment at 4°C or room temperature for 6 months, resulting in oocysts sporulating and living for an extended period of time [40]. Oocysts are resistant to freezing [41] but are destroyed by prolonged desiccation [42], high temperatures (60°C), and exposure to UV light [43]. Therefore, any Env parameters, DH, and IH influence the completion of *T. gondii* life cycle. Since cats only shed it once, seropositive cats no longer shed *T. gondii* after their first infection; the presence of IHs encourages the transmission of parasites throughout an environment if the IHs provide a scarce source of food for the DH.

1.4 *T. gondii* in humans and animals

Humans, as IH, present a third or more people who have been exposed to infection with *T. gondii*. The prevalence ranges widely (10–85%) between nations, geographical areas within a country [7], socioeconomic conditions, climate (warm, humid tropical areas, cold-temperate, or desert regions), poor sanitation, a lack of clean water, and poor hygiene [44]. Toxoplasmosis is classified into an acute acquired infection (horizontal transmission) and a congenital infection (CT) [45] with an incubation period from 10 to 23 days after consuming tissue cysts and from 5 to 10 days following the consumption of oocysts [10, 46]. CT to the fetus (transplacental) may occur if the mother gets a primary infection or the chronically infected mother becomes immunocompromised during pregnancy [13]. Also, blood transfusion, solid organ transplant, or bone marrow transplant from a seropositive donor might be considered for infection [47]. Sexual transmission is established in many animal species, and circumstantial evidence implies that it may also occur in humans [48–50]. Because most infections are subclinical or misdiagnosed, it is challenging to establish the human toxoplasmosis and welfare effects [42]. Moderate flu-like symptoms, such as fever, muscle discomfort, and fatigue, may occur in some people [51]. Cases of severe toxoplasmosis and ocular disease are documented [52]. The most common cause is thought to be consuming raw or undercooked meat from an infected animal [53]. Unpasteurized milk; contaminated meat, vegetables, and fruits; and contaminated water might be a source of infection [54]. Infection may be caused by consuming water or food contaminated with feline excrement [55]. Disease transmission can be prevented by adequately boiling or freezing; temperatures over 67°C or below −12°C render tissue cysts nonviable [56, 57], and salted and cured meat often weakens the viability. However, *Toxoplasma* tissue cysts seem refractory to these approaches [58]. The efficiency of thermal, nonthermal, and chemical/biochemical treatments for inactivating *T. gondii* in foods destined for human consumption has been reviewed [59]. When these individuals have a healthy immune system, the infection may not cause any symptoms. However, if the immune system is suppressed, the parasite can reactivate and cause potentially fatal complications [60]. Clinical signs of the disease include encephalitis, schizophrenia, bipolar disorders, depression, obsessive-compulsive disorders, retinochoroiditis, myocarditis, and fetal abnormalities following transplacental infection in immunocompromised people [61]. Both immunocompetent and immunocompromised pregnant women are at risk of transmitting *T. gondii* to their unborn children if they become infected with the parasite during their pregnancies [62]. The prevalence of CT varies by nation and location. Pregnancy-related fever, eating unwashed vegetables, and diagnosis in the third trimester were all linked to CT [63]. It is believed that the high genetic diversity of *T. gondii* seen in animals from this country correlates with illness severity. Infections after birth are usually undetectable, and toxoplasmosis predominantly affects immunocompromised patients where the clinical course of the illness is frequently considerably more severe [51]. Infected newborns might have nonspecific symptoms [64]. However, most infections are asymptomatic because the organism has developed several strategies to prevent significant inflammatory and immune responses. In the case of infections in wildlife populations, *T. gondii* infects any warm-blooded wildlife by ingesting environmental oocysts or, in omnivores and carnivores, both oocysts and tissue cysts [65, 66]. Production species are also affected by *T. gondii*, such as poultry [67]. Like chickens, *T. gondii* is the IH in livestock [68, 69], sheep [70, 71], and swine [72]. During infection of IH, first contact is made when parasites infiltrate enterocytes or cross the epithelial barrier to infect resident macrophages, dendritic cells (DCs), and lymphocytes [73] and rapidly multiply, infecting any cell with a nucleus; the resulting cell lysis may trigger rapid disease onset. A solid first line of defense is encouraged by the innate immune response. Within a few days of infection, the body develops an adaptive immune response that limits the infection and strengthens resistance to the disease. In response, the parasite undergoes a transition into the bradyzoite, a latent stage that develops inside the parasite cysts within the host. The immune system cannot combat these tissue cysts [16]. The complex relationship between *T. gondii* and innate immune system cells has been the subject of several researches and serves as a model for various microbial illnesses [74].

1.5 Conclusion

Since the first isolation over a century ago, *Toxoplasma gondii* has been extensively studied to understand its biology, transmission, life cycle, and implications for the health of its definitive hosts and intermediate hosts, including the human being. The life cycle of this protozoan of the phylum Apicomplexa presents different stages (sexual and asexual), which, in combination with a capacity for resistance to the environment and multiple hosts, makes this organism highly adaptable and easy to transmit and spread. Sequencing of *T. gondii* genome allowed the development of databases applied to the study, diagnosis, and typification, as well as the analysis of pathogenic factors associated with the capacity. Particularly, *T. gondii* is mainly mentioned in the literature due to its potential effect on various species, including humans. The DH-environment-IH interactions form the essential ecological components for its dissemination, mainly by fecal material/oral cycle. Human infection is highly impacted worldwide since it is presumed that more than a third of the human population has been infected by this protozoan. Due to the immune system being actively involved in controlling infection, clinical signs range from a simple cold to severe illness in the case of an immunodeficient situation (eye lesions, lymphopathy, embryonic damage, and mental disorders). Its study demonstrated its presence in many foods of animal and vegetable origin, placing this infection as one of the main risks of food-borne diseases. Its presence in wildlife is also an intrinsically related factor in its worldwide distribution. Due to this, *T. gondii* represents a high risk for humans and a great variety of IHs that can be affected by this parasite; therefore, the study is highly relevant for its control and research worldwide.

Author details

Saeed El-Ashram[1,2]*, Gungor Cagdas Dincel[3], Mohammed Hamdy Farouk[4], Abdulaziz Alouffi[5], Hany M.R. Abdel-Latif[6], Shagufta Iqbal[7], Guillermo Tellez-Isaias[8], Billy M. Hargis[9], Víctor Manuel Petrone-García[10], Gabriela Aguilar-Tipacamú[11] and Inkar Castellanos-Huerta[12]

1 Faculty of Science, Kafrelsheikh University, Kafr El-Sheikh, Egypt

2 College of Life Science and Engineering, Foshan University, Foshan, China

3 Eskil Vocational School, Laboratory and Veterinary Science, Aksaray University, Aksaray, Turkey

4 Animal Production Department, Faculty of Agriculture, Al-Azhar University, Nasr City, Cairo, Egypt

5 King Abdulaziz City for Science and Technology, Riyadh, Saudi Arabia

6 Department of Poultry and Fish Diseases, Faculty of Veterinary Medicine, Alexandria University, Alexandria, Egypt

7 Department of Zoology, University of Kashmir, India

8 Department of Poultry Science, University of Arkansas, Fayetteville, USA

9 Department of Poultry Science, University of Arkansas Agricultural Experiment Station, Fayetteville, United States

10 Petrone-García Departamento de Ciencias Pecuarias, Facultad de Estudios Superiores Cuautitlán UNAM, Cuautitlán, Mexico

11 Licenciatura en Medicina Veterinaria y Zootecnia, Facultad de Ciencias Naturales, Universidad Autónoma de Querétaro, Querétaro, México

12 Programa de Maestría y Doctorado en Ciencias de la Producción y de la Salud Animal, Facultad de Medicina Veterinaria y Zootecnia, Universidad Nacional Autónoma de México, Ciudad Universitaria, Ciudad de Mexico, Mexico

*Address all correspondence to: saeed_elashram@yahoo.com

IntechOpen

© 2023 The Author(s). Licensee IntechOpen. This chapter is distributed under the terms of the Creative Commons Attribution License (http://creativecommons.org/licenses/by/3.0), which permits unrestricted use, distribution, and reproduction in any medium, provided the original work is properly cited. (cc) BY

References

[1] Dubey J. History of the discovery of the life cycle of toxoplasma gondii. International Journal for Parasitology. 2009;**39**:877-882

[2] Sharifpour MF. Artificial regulation of transcription in toxoplasma gondii, the model apicomplexan [PhD thesis]. 2021

[3] Dubey J. Advances in the life cycle of toxoplasma gondii. International Journal for Parasitology. 1998;**28**:1019-1024

[4] Weiss LM. Toxoplasma Gondii: The Model Apicomplexan-Perspectives and Methods. 2nd ed. Amsterdam, Netherlands: Elsevier's Science and Technology; 2013

[5] Yan C, Liang L-J, Zheng K-Y, Zhu X-Q. Impact of environmental factors on the emergence, transmission and distribution of toxoplasma gondii. Parasites & Vectors. 2016;**9**:1-7

[6] Dumètre A, Dardé M-L. How to detect toxoplasma gondii oocysts in environmental samples? FEMS Microbiology Reviews. 2003;**27**:651-661

[7] Robert-Gangneux F, Dardé M-L. Epidemiology of and diagnostic strategies for toxoplasmosis. Clinical Microbiology Reviews. 2012;**25**:264-296

[8] Ferguson DJ, Dubremetz J-F. The ultrastructure of toxoplasma gondii. Toxoplasma Gondii. 2014:19-59

[9] Kieffer F, Wallon M. Congenital toxoplasmosis. Handbook of Clinical Neurology. 2013;**112**:1099-1101

[10] Hampton MM. Congenital toxoplasmosis: A review. Neonatal Network. 2015;**34**:274-278

[11] Simon A, Poulin MB, Rousseau AN, Ogden NH. Fate and transport of toxoplasma gondii oocysts in seasonally snow covered watersheds: A conceptual framework from a melting snowpack to the Canadian arctic coasts. International Journal of Environmental Research and Public Health. 2013;**10**:994-1005

[12] Muñoz-Zanzi CA, Fry P, Lesina B, Hill D. Toxoplasma gondii oocyst–specific antibodies and source of infection. Emerging Infectious Diseases. 2010;**16**:1591

[13] Harker K, Ueno N, Lodoen M. Toxoplasma gondii dissemination: A parasite's journey through the infected host. Parasite Immunology. 2015;**37**:141-149

[14] Randall L, Hunter C. Parasite dissemination and the pathogenesis of toxoplasmosis. European Journal of Microbiology & Immunology. 2011;**1**:3-9

[15] Tomita T, Bzik DJ, Ma YF, Fox BA, Markillie LM, Taylor RC, et al. The toxoplasma gondii cyst wall protein CST1 is critical for cyst wall integrity and promotes bradyzoite persistence. PLoS Pathogens. 2013;**9**:e1003823

[16] Sullivan WJ Jr, Jeffers V. Mechanisms of toxoplasma gondii persistence and latency. FEMS Microbiology Reviews. 2012;**36**:717-733

[17] Walochnik J, Duchêne M. Molecular Parasitology: Protozoan Parasites and Their Molecules. Heidelberger Platz 3, Berlin: Springer; 2016

[18] Khan A, Taylor S, Su C, Sibley LD, Paulsen I, Ajioka JW. Genetics and genome organization of toxoplasma gondii. In: Toxoplasma: Molecular and Cellular Biology. Norfolk, England: Horizon Bioscience; 2007. pp. 193-207

[19] Roos DS. Themes and variations in apicomplexan parasite biology. Science. 2005;**309**:72-73

[20] Gajria B, Bahl A, Brestelli J, Dommer J, Fischer S, Gao X, et al. ToxoDB: An integrated toxoplasma gondii database resource. Nucleic Acids Research. 2007;**36**:D553-D556

[21] Sibley LD, Boothroyd JC. Virulent strains of toxoplasma gondii comprise a single clonal lineage. Nature. 1992;**359**:82-85

[22] Ajioka JW, Boothroyd JC, Brunk BP, Hehl A, Hillier L, Manger ID, et al. Gene discovery by EST sequencing in toxoplasma gondii reveals sequences restricted to the apicomplexa. Genome Research. 1998;**8**:18-28

[23] Galal L, Ariey F, Gouilh MA, Dardé M-L, Hamidović A, Letourneur F, et al. A unique toxoplasma gondii haplotype accompanied the global expansion of cats. Nature Communications. 2022;**13**:5778

[24] Saeij JP, Boyle JP, Boothroyd JC. Differences among the three major strains of toxoplasma gondii and their specific interactions with the infected host. Trends in Parasitology. 2005;**21**:476-481

[25] Frazão-Teixeira E, Sundar N, Dubey J, Grigg M, De Oliveira F. Multi-locus DNA sequencing of toxoplasma gondii isolated from Brazilian pigs identifies genetically divergent strains. Veterinary Parasitology. 2011;**175**:33-39

[26] Vaudaux JD, Muccioli C, James ER, Silveira C, Magargal SL, Jung C, et al. Identification of an atypical strain of toxoplasma gondii as the cause of a waterborne outbreak of toxoplasmosis in Santa Isabel do Ivai, Brazil. The Journal of Infectious Diseases. 2010;**202**:1226-1233

[27] Grigg ME, Suzuki Y. Sexual recombination and clonal evolution of virulence in toxoplasma. Microbes and Infection. 2003;**5**:685-690

[28] Ajzenberg D, Yera H, Marty P, Paris L, Dalle F, Menotti J, et al. Genotype of 88 toxoplasma gondii isolates associated with toxoplasmosis in immunocompromised patients and correlation with clinical findings. The Journal of Infectious Diseases. 2009;**199**:1155-1167

[29] Mercier A, Devillard S, Ngoubangoye B, Bonnabau H, Bañuls A-L, Durand P, et al. Additional haplogroups of toxoplasma gondii out of Africa: Population structure and mouse-virulence of strains from Gabon. PLoS Neglected Tropical Diseases. 2010;**4**:e876

[30] Suzuki Y, Orellana MA, Schreiber RD, Remington JS. Interferon-γ: The major mediator of resistance against toxoplasma gondii. Science. 1988;**240**:516-518

[31] Howe DK, Sibley LD. Toxoplasma gondii comprises three clonal lineages: Correlation of parasite genotype with human disease. The Journal of Infectious Diseases. 1995;**172**:1561-1566

[32] Su C, Khan A, Zhou P, Majumdar D, Ajzenberg D, Dardé M-L, et al. Globally diverse toxoplasma gondii isolates comprise six major clades originating from a small number of distinct ancestral lineages. Proceedings of the National Academy of Sciences. 2012;**109**:5844-5849

[33] Silva NM, Rodrigues CV, Santoro MM, Reis LF, Alvarez-Leite JI, Gazzinelli RT. Expression of indoleamine 2, 3-dioxygenase, tryptophan

degradation, and kynurenine formation during in vivo infection with toxoplasma gondii: Induction by endogenous gamma interferon and requirement of interferon regulatory factor 1. Infection and Immunity. 2002;**70**:859-868

[34] Niedelman W, Gold DA, Rosowski EE, Sprokholt JK, Lim D, Farid Arenas A, et al. The rhoptry proteins ROP18 and ROP5 mediate toxoplasma gondii evasion of the murine, but not the human, interferon-gamma response. PLoS Pathogens. 2012;**8**:e1002784

[35] Etheridge RD, Alaganan A, Tang K, Lou HJ, Turk BE, Sibley LD. The toxoplasma pseudokinase ROP5 forms complexes with ROP18 and ROP17 kinases that synergize to control acute virulence in mice. Cell Host & Microbe. 2014;**15**:537-550

[36] Steinfeldt T, Könen-Waisman S, Tong L, Pawlowski N, Lamkemeyer T, Sibley LD, et al. Phosphorylation of mouse immunity-related GTPase (IRG) resistance proteins is an evasion strategy for virulent toxoplasma gondii. PLoS Biology. 2010;**8**:e1000576

[37] Taylor MA, Coop RL, Wall RL. Veterinary Parasitology. New York, NY: John Wiley & Sons; 2015

[38] Doherty TS, Dickman CR, Johnson CN, Legge SM, Ritchie EG, Woinarski JC. Impacts and management of feral cats felis catus in Australia. Mammal Review. 2017;**47**:83-97

[39] Frenkel J, Ruiz A, Chinchilla M. Soil survival of toxoplasma oocysts in Kansas and Costa Rica. The American Journal of Tropical Medicine and Hygiene. 1975;**24**:439-443

[40] Lindsay DS, Collins MV, Mitchell SM, Cole RA, Flick GJ, Wetch CN, et al. Sporulation and survival of toxoplasma gondii oocysts in seawater. The Journal of Eukaryotic Microbiology. 2003;**50**:687-688

[41] Frenkel J, Dubey J, et al. Effects of freezing on the viability of toxoplasma oocysts. The Journal of Parasitology. 1973;**59**:587-588

[42] Dubey JP. Toxoplasmosis of Animals and Humans. Boca Raton, Florida, USA: CRC Press; 2016

[43] Dumètre A, Le Bras C, Baffet M, Meneceur P, Dubey J, Derouin F, et al. Effects of ozone and ultraviolet radiation treatments on the infectivity of toxoplasma gondii oocysts. Veterinary Parasitology. 2008;**153**:209-213

[44] Flegr J, Prandota J, Sovičková M, Israili ZH. Toxoplasmosis–A global threat. Correlation of latent toxoplasmosis with specific disease burden in a set of 88 countries. PLoS One. 2014;**9**:e90203

[45] Hussain MA, Stitt V, Szabo EA, Nelan B. Toxoplasma gondii in the food supply. Pathogens. 2017;**6**:21

[46] Dubey J, Murata F, Cerqueira-Cézar C, Kwok O, Villena I. Congenital toxoplasmosis in humans: An update of worldwide rate of congenital infections. Parasitology. 2021;**148**:1406-1416

[47] Singh S. Congenital toxoplasmosis: Clinical features, outcomes, treatment, and prevention. Tropical Parasitology. 2016;**6**:113

[48] Liu S, Qin C, Yao Z, Wang D. Study on the transmission of toxoplasma gondii by semen in rabbits. Zhongguo Ji Sheng Chong Xue Yu Ji Sheng Chong Bing Za Zhi. 2006;**24**:166-170

[49] Hlaváčová J, Flegr J, Řežábek K, Calda P, Kaňková Š. Male-to-female

presumed transmission of toxoplasmosis between sexual partners. American Journal of Epidemiology. 2021;**190**:386-392

[50] Arantes TP, Lopes WDZ, Ferreira RM, Pieroni JSP, Pinto VM, Sakamoto CA, et al. Toxoplasma gondii: Evidence for the transmission by semen in dogs. Experimental Parasitology. 2009;**123**:190-194

[51] Jg M, Liesenfeld O. Toxoplasmosis. Lancet. 2004;**363**:1965-1976

[52] Demar M, Hommel D, Djossou F, Peneau C, Boukhari R, Louvel D, et al. Acutetoxoplasmosesinimmunocompetent patients hospitalized in an intensive care unit in French Guiana. Clinical Microbiology and Infection. 2012;**18**:E221-E231

[53] Sobanski V, Ajzenberg D, Delhaes L, Bautin N, Just N. Severe toxoplasmosis in immunocompetent hosts: Be aware of atypical strains. American Journal of Respiratory and Critical Care Medicine. 2013;**187**:1143-1145

[54] Hill DE, Dubey JP. Toxoplasma Gondii as a Parasite in Food: Analysis and Control. Microbiol Spectrum 4: PFS-0011-2015. DOI: 10.1128/Microbiolspec. PFS-0011-2015. 2016

[55] Organization, WH. Foodborne Disease Outbreaks: Guidelines for Investigation and Control. Geneva, Switzerland: World Health Organization; 2008

[56] Kotula A, Dubey J, Sharar A, Andrews C, Shen S, Lindsay D. Effect of freezing on infectivity of toxoplasma gondii tissue cysts in Pork. Journal of Food Protection. 1991;**54**:687-690

[57] Dubey J, Kotula A, Sharar A, Andrews C, Lindsay D. Effect of high temperature on infectivity of toxoplasma gondii tissue cysts in Pork. The Journal of Parasitology. 1990;**76**(2):201-204

[58] Pott S, Koethe M, Bangoura B, Zöller B, Daugschies A, Straubinger R, et al. Effects of PH, sodium chloride, and curing salt on the infectivity of toxoplasma gondii tissue cysts. Journal of Food Protection. 2013;**76**:1056-1061

[59] Mirza Alizadeh A, Jazaeri S, Shemshadi B, Hashempour-Baltork F, Sarlak Z, Pilevar Z, et al. A review on inactivation methods of toxoplasma gondii in foods. Pathogens and Global Health. 2018;**112**:306-319

[60] Porter SB, Sande MA. Toxoplasmosis of the central nervous system in the acquired immunodeficiency syndrome. The New England Journal of Medicine. 1992;**327**:1643-1648

[61] Daher D, Shaghlil A, Sobh E, Hamie M, Hassan ME, Moumneh MB, et al. Comprehensive overview of toxoplasma gondii-induced and associated diseases. Pathogens. 2021;**10**:1351

[62] Dunn D, Wallon M, Peyron F, Petersen E, Peckham C, Gilbert R. Mother-to-child transmission of toxoplasmosis: Risk estimates for clinical counselling. The Lancet. 1999;**353**:1829-1833

[63] Vivacqua DPF, Paz AB, Frota ACC, Penna CRR, Martins MG, Abreu TF, et al. Antenatal factors related to congenital toxoplasmosis in Rio De Janeiro, Brazil. Journal of Maternal-fetal and Neonatal Medicine. 2022;**35**:7200-7206

[64] Andreoletti O, Budka H, Buncic S, Colin P, Collins JD, De A, et al. Surveillance and monitoring of toxoplasma in humans, food and animals scientific opinion of the panel on biological hazards. EFSA Journal. 2007;**583**:1-64

[65] Lopes AP, Sargo R, Rodrigues M, Cardoso L. High seroprevalence of

antibodies to toxoplasma gondii in wild animals from Portugal. Parasitology Research. 2011;**108**:1163-1169

[66] Wendte JM, Gibson AK, Grigg ME. Population genetics of toxoplasma gondii: New perspectives from parasite genotypes in wildlife. Veterinary Parasitology. 2011;**182**:96-111

[67] Chumpolbanchorn K, Lymbery A, Pallant L, Pan S, Sukthana Y, Thompson R. A high prevalence of toxoplasma in Australian chickens. Veterinary Parasitology. 2013;**196**:209-211

[68] Dubey J. Toxoplasma gondii infections in chickens (gallus domesticus): Prevalence, clinical disease, diagnosis and public health significance. Zoonoses and Public Health. 2010;**57**:60-73

[69] van der Giessen J, Fonville M, Bouwknegt M, Langelaar M, Vollema A. Seroprevalence of trichinella spiralis and toxoplasma gondii in pigs from different housing systems in The Netherlands. Veterinary Parasitology. 2007;**148**:371-374

[70] Vesco G, Buffolano W, La Chiusa S, Mancuso G, Caracappa S, Chianca A, et al. Toxoplasma gondii infections in sheep in Sicily, Southern Italy. Veterinary Parasitology. 2007;**146**:3-8

[71] Innes EA, Bartley PM, Buxton D, Katzer F. Ovine toxoplasmosis. Parasitology. 2009;**136**:1887-1894

[72] Hill D, Dubey J, Baroch J, Swafford S, Fournet V, Hawkins-Cooper D, et al. Surveillance of feral swine for Trichinella spp. and Toxoplasma gondii in the USA and host-related factors associated with infection. Veterinary Parasitology. 2014;**205**:653-665

[73] Buzoni-Gatel D, Schulthess J, Menard LC, Kasper LH. Mucosal defences against orally acquired protozoan parasites, emphasis on toxoplasma gondii infections. Cellular Microbiology. 2006;**8**:535-544

[74] David Sibley L. Invasion and intracellular survival by protozoan parasites. Immunological Reviews. 2011;**240**:72-91

Chapter 2

Development of Schizont Stages of *Toxoplasma gondii* in Primary Cell Culture of Feline Enterocytes

Renata M. de Muno, Marcos A. Moura, Letícia C. Medeiros, Pedro N. Caldas, Rafael M. Mariante and Helene S. Barbosa

Abstract

Intestinal epithelial cell cultures are a potentially applicable model for investigating enteropathogens such as the protozoan *Toxoplasma gondii*, the etiological agent of toxoplasmosis. Felids such as domestic cats are the only known definitive hosts where the parasite undergoes sexual reproduction, which occurs in the enterocytes. Primary feline intestinal epithelial cell (FIEC) cultures were obtained from the fetal small gut of felines, and the epithelial nature of these cells was confirmed by the revelation of cytokeratin and intestinal alkaline phosphatase content by fluorescence microscopy, besides alignment, microvilli, and adherent intercellular junctions by ultrastructural analysis. FIECs infected with *T. gondii* bradyzoite forms showed that the parasite:cell ratio was determinant for establishing the lytic cycle and cystogenesis and the induction of schizont-like forms. Type C and D schizonts were identified by light and electron microscopies, which showed morphological characteristics like those previously described based on the analysis of cat intestines experimentally infected with *T. gondii*. These data indicate that FIECs simulate the microenvironment of the felid intestine, allowing the development of schizogony and classic endopolygeny. This cellular framework opens new perspectives for the *in vitro* investigation of biological and molecular aspects involved in the *T. gondii* enteric cycle.

Keywords: *Toxoplasma gondii*, enteric cycle, primary cell culture, feline enterocytes, schizont stages

1. Introduction

Intestinal cells act as barriers to prevent access to potentially harmful substances and the migration of the underlying cells in the lamina propria [1]. Several investigators have established culture methods for intestinal cells from different animal species that mimic normal intestinal development [2]. Culture techniques have been developed for cells from different sources, including adult [3, 4] and embryonic cells [5–8]. The introduction of growth factors or interaction of these systems with the extracellular matrix during recent decades has allowed the development of experimental approaches to study *in vitro* differentiation [9, 10]. These advances have afforded the

 IntechOpen

possible application of enterocyte cultures for *in vitro* studies (e.g., the interaction of these cells with enteroparasites [11] such as *Toxoplasma gondii* [8]).

T. gondii, the etiological agent of toxoplasmosis, is an obligatory intracellular parasite that causes one of the most common zoonoses in the world. Its transmission occurs by (i) oral infection via the ingestion of cysts, present in raw or poorly cooked meat; (ii) ingestion of oocysts, present in the feces of Felidae contaminating food, water, and soil [12, 13]; or (iii) vertical transmission via the transplacental route [14, 15]. Cats and other felines are the only definitive hosts capable of directly spreading *T. gondii* in the environment, since the enteroepithelial cycle, including the sexual stage of the parasite, occurs exclusively in those species [16]. The infection in felids is established after ingestion of cysts or oocysts in tissues whose walls are destroyed by proteolytic enzymes in the stomach and intestine, resulting in the release of bradyzoites or sporozoites that invade intestinal cells and initiate the enteroepithelial cycle of the parasite [17].

Five distinct enteroepithelial morphological stages or schizonts of *T. gondii* (Types: A, B, C, D, and E) are described in the felid gut, involving the processes of schizogony, gametogony, and sporogony, which result in the formation of immature oocysts [16–19]. Specific knowledge about the enteric cycle of *T. gondii* in Felidae is limited to morphological characterization *in vivo* [17, 20–25], which makes it difficult to monitor the kinetics of infection. Thus, the analysis of the temporal events established during this *in vivo* cycle is subjective since histological studies of the gut do not allow monitoring the actual sequence of differentiation events of the parasite's infectious stages. However, the practice of euthanasia of cats for scientific studies is restricted [26], which corroborates the need to introduce alternative research models to explore the enteroepithelial cycle of *T. gondii* in felids.

No cell models of the feline intestinal epithelium are commercially available to allow the study of the *T. gondii* enteroepithelial cycle *in vitro.* Several attempts to culture intestinal epithelial cells from adult animals or establish normal cell lines derived from normal enterocytes have not been very successful [27]. The lack of cellular models that allow the reproduction of the enteric cycle of *T. gondii* in felids motivated us to introduce primary cultures of cat enterocytes as an alternative for this study. Preliminary data from this interaction have been published by our group [8]. We now deepen this study by revealing the *T. gondii* development *in vitro* in feline enterocytes.

2. Experimental design

2.1 Feline enterocyte primary cell culture

Feline enterocyte primary cell cultures (FIEC) were obtained from fetuses of a clinically healthy pregnant domestic cat (no gastrointestinal disease and serologically negative for *T. gondii*, feline immunodeficiency virus, and feline leukemia virus). All procedures were performed in accordance with the guidelines stipulated by the Brazilian College of Animal Experimentation (COBEA). This study was approved by the Fundação Oswaldo Cruz Committee of Ethics for the Use of Animals (license L042/2018 A1).

Small intestine samples corresponding to the jejunum-ileum region (~5 cm) were collected aseptically. The samples were dissected, and the fragments were gathered in ice-cold sterile phosphate-buffered saline (PBS) with a 10% antibiotic solution (Sigma-Aldrich-St. Louis, MO, United States). This tissue was opened longitudinally, washed three times with PBS, and maintained in this solution with 10% antibiotics for 20 min at room temperature. Fragments were divided into small pieces (1 cm^3) and washed into PBS. The fragments were placed in nonenzymatic dissociation buffer (pH 7.2) containing 1 mM EDTA (Sigma-Aldrich-St. Louis, MO, United States), 1 mM EGTA (Sigma-Aldrich-St. Louis, MO, United States), 0.5 mM dithiothreitol (Sigma-Aldrich-St. Louis, MO, United States), and 10% antibiotic solution for 20 min under stirring at room temperature [2–6]. The cell aggregates were plated in DMEM/Hams medium Dulbecco's Modified Eagle's Medium/Ham's Nutrient F12 (1:1) containing 1% antibiotic solution, 1 mM glutamine, 5% fetal bovine serum (Life Technologies, São Paulo, SP, Brazil), 20 ng/ml epidermal growth factor (Sigma-Aldrich-St. Louis, MO, United States) [4, 7], 0.1% human insulin (Humulin N - Lilly, Indianapolis, IN, United States), 100 nM hydrocortisone (Sigma-Aldrich-St. Louis, MO, United States), 1% nonessential amino acids 100x (Life Technologies, São Paulo, SP, Brazil), and 1 µg/ml 3,3′, 5-triiodo-L-thyronine sodium salt (Sigma-Aldrich-St. Louis, MO, United States) [10]. The cultures were maintained at 37°C in a 5% CO_2 atmosphere, and the medium was renewed every two days.

Confluent FIECs were treated for 10 min at 37°C with dissociation solution (PBS with 0.01% EDTA and 0.25% trypsin). After dissociation, the cells were placed in culture medium at 4°C with 10% fetal bovine serum to inhibit the action of trypsin, centrifuged for 7 min at 650 × *g* at 4°C, and grown in 24-well plates on coverslips (10^5 cells/well) or on 35 mm^3 plastic disks (5.0 x 10^5 cells/disk). Cell cultures were daily analyzed by light microscopy (Zeiss Imager A2 microscope) after fixation in Bouin's solution and Giemsa staining to assess morphology and proliferation. Images were captured with Soft Axion Vision 40 v.4.8.2.0 and an Axion cam MRc color camera.

2.2 Characterization of FIEC by immunolabeling

Several monoclonal antibodies were applied to characterize the FIECs: anti-pancytokeratin clone PCK-26 (Sigma-Aldrich-St. Louis, MO, United States); anti-vimentin clone VIM-13.2 (Sigma-Aldrich-St. Louis, MO, United States); anti-intestinal alkaline phosphatase clone AP-59 (Sigma-Aldrich-St. Louis, MO, United States); and anti-desmin clone DE-U-10 (Sigma-Aldrich-St. Louis, MO, United States). The cells were fixed for 10 min at 4°C with 4% paraformaldehyde in PBS, washed three times for 10 min in PBS, and then incubated for 30 min in 50 mM ammonium chloride to block free aldehyde radicals. Afterward, the cells were permeabilized for 20 min in a PBS solution containing 0.05% Triton X-100 (Roche, Rio de Janeiro, RJ, Brazil) and 4% BSA (Sigma-Aldrich-St. Louis, MO, United States) to block nonspecific binding. For the indirect immunofluorescence assays, the cells were incubated for 2 h at 37°C with the following primary antibodies: anti-vimentin (1:200); anti-cytokeratin (1:100); anti-intestinal alkaline phosphatase; and anti-desmin (1:100). After this incubation, the cells were washed 3 times for 10 min in PBS containing 4% BSA and incubated for 1 h at 37°C with the secondary antibody at a dilution of 1:1000 (mouse anti-IgG conjugated with FITC or TRITC) (Sigma-Aldrich-St. Louis, MO, United States). To reveal actin filaments, the cells were incubated for 1 h at 37°C with 4 µg/mL phalloidin-FITC in PBS (Sigma-Aldrich-St. Louis, MO, United States). Next, the cultures were washed 3 times for 10 min in PBS and incubated for 5 min with 0.1 µg/mL 4′,6-diamidino-2-phenylindole (DAPI, Sigma-Aldrich-St. Louis, MO, United States) diluted in PBS. After wash, the coverslips were mounted on slides with a solution of 2.5% DABCO (1,4-diazabicyclo-[2,2,2]-octane-triethylenediamine)

(Sigma-Aldrich-St. Louis, MO, United States) in PBS containing 50% glycerol, pH 7.2. Controls were performed by omission of the primary antibody. The samples were examined with a confocal laser-scanning microscope (CLSM Axiovert 510, META, Zeiss) with a 543 helium laser (LP560 filter), 488 argon/krypton laser (Ar/Kr) (filter LP515), and a 405 Diode laser (LP 420 filter).

2.3 Isolation of *T. gondii* bradyzoites and interaction with FIEC

T. gondii cysts from the ME-49 strain (Type II) previously isolated from infected mice were inoculated intraperitoneally into female C57BL/6 mice (15–18 g) with 50 cysts/animal. Four and 12 weeks after infection, the mice were sacrificed and the brain cysts isolated, as described by Freyre [28] and modified by Guimarães et al. [29]. In addition, bradyzoites were obtained from isolated tissue cysts according to Popiel et al. [30].

Confluent FIEC cultures were infected with *T. gondii* bradyzoites. The assays were performed at 1:5, 1:10, and 1:20 (parasite: host cell) ratios for periods ranging from 1 to 9 days postinfection to analyze the infection course and to evaluate the parasite's intracellular fate. After different periods of interaction, the cells were washed in PBS and processed according to the experiments to be performed.

The ability of the *T. gondii* bradyzoites to infect FIECs *in vitro* was examined by fixation of the cells in Bouin's solution and Giemsa staining. The infection analysis was carried out with 200 cells per coverslip in three independent experiments at interaction times of 24, 48, 72, and 96 hours, in duplicate. Data were analyzed in GraphPad Prism 7.0 using Two-Way ANOVA with Tukey's multiple comparisons test, and results were expressed as mean with SEM (graph) or SD (table). Differences were considered statistically significant when the p-value was <0.05.

2.4 Characterization of *T. gondii* stages by immunolabeling

The differentiation of parasites in culture cells infected with bradyzoite forms of *T. gondii* was monitored using tachyzoite stage-specific anti-SAG-1 antibodies (kindly provided by Dr. José Roberto Mineo - Immunoparasitology Laboratory, Federal University of Uberlândia, Minas Gerais, Brazil). Bradyzoites of *T. gondii* were identified with stage-specific anti-BAG-1 antibodies (anti-BAG1-7E5; kindly supplied by Dr. Wolfgang Bohne - Institut für Medizinische Mikrobiologie, Universität Göttingen, Germany) as previously described [31]. Initially, the cultures were fixed for 20 min at 4°C on days 1 to 4 with 4% PFA in PBS, washed three times for 10 min in PBS, and then incubated for 30 min in 50 mM ammonium chloride to block free aldehyde radicals. Next, the cells were permeabilized for 20 min in a PBS solution containing 0.05% Triton X-100 (Roche, Rio de Janeiro, RJ, Brazil) and 4% BSA (Sigma-Aldrich-St. Louis, MO, United States) to block nonspecific binding. For the indirect immunofluorescence assay, the host cells were incubated for 2 h at 37°C with the primary antibodies anti-SAG-1 (1:200) and anti-BAG (1:500) diluted in PBS/BSA. After incubation, the cells were washed with PBS containing 4% BSA and incubated for 1 h at 37°C with the secondary antibody at a 1:1000 dilution (anti-mouse IgG conjugated with FITC). Controls were performed by omission of the primary antibodies. Afterward, the cultures were washed 3 times for 10 min in PBS and processed for fluorescence microscopy as described above.

2.5 Ultrastructural analysis

FIECs infected or not with bradyzoite forms of *T. gondii* were washed 3 times for 10 min with PBS and fixed for 1 h at 4°C in 2.5% glutaraldehyde diluted in a 0.1 M sodium cacodylate buffer containing 3.5% sucrose and 2.5 mM $CaCl_2$ (pH 7.2). After fixation, the cells were washed in the same buffer and then post-fixed for 30 min at room temperature in 1% osmium tetroxide diluted in a 0.1 M Na-cacodylate buffer. For transmission electron microscopy analysis, the cells were washed in the same buffer, scraped from the plastic dish at 4°C, and centrifuged. Then, the cells were dehydrated in a graded acetone series and embedded in an epoxy resin (PolyBed 812). Thin sections were stained with uranyl acetate and lead citrate and then examined under a transmission electron microscope (Jeol JEM1011). For scanning electron microscopy, the FIECs were fixed for 30 min at room temperature with 2.5% glutaraldehyde in 0.1 M Na-cacodylate buffer (pH 7.2) and post-fixed for 30 min at room temperature with a solution of 1% OsO4 containing 2.5 mM $CaCl_2$ in the same buffer. The cells were dehydrated in an ascending acetone series and dried by the critical point method with CO_2 (CPD 030, Balzers, Switzerland). The samples were mounted on aluminum stubs, coated with a 20 nm layer of gold (Cressington Sputter Coater 108), and examined under a scanning electron microscope (Jeol JSM 6390LV) at the Rudolf Barth Electron Microscopy Platform at Oswaldo Cruz Institute.

3. Results

3.1 Morphological characteristics of feline intestinal cells *in vitro*

The attachment of FIECs to the substrate was observed by phase-contrast microscopy, in situ. In 5 days, the cells aligned and polarized, with the nuclei located in the same plane as the organization of the columnar epithelium (**Figure 1A**). Ultrastructural analysis by scanning electron microscopy showed the absorptive characteristics of these cells, including the identification of plasma membrane projections that established focal adhesion points (**Figure 1B**). Long and thin finger-like projections were visible and often established cell-cell contacts and extensive cytoplasmic contacts, indicating the formation of specialized membrane areas, such as cellular junctions (**Figure 1B**). Transmission electron microscopy demonstrated that epithelial cells in culture retained a great number of cytological features typical of intestinal epithelial cells, such as large numbers of microvilli (i.e., a brush border at the apical pole) (**Figure 1C–D**). Lateral interdigitations are observed below the junctional complex between two adjacent epithelial cells (**Figure 1C**). The junctional areas presented tight junctions (zonula occludens) in which the outer leaflets of the plasma membranes were fused, intermediate junctions (zonula adherens), characterized by plasma membranes separated by a space, and desmosomes (macula adherens) (**Figure 1D**). All these characteristics confirmed the intestinal epithelial nature of FIECs as enterocytes that were maintained for up to six passages.

3.2 Expression of intestinal markers in FIEC

To confirm the epithelial nature of FIEC, we investigated the intermediate filaments by employing an anti-pan-cytokeratin antibody that recognizes a range of cytokeratins (1, 5, 6, 8, and 10). Confocal laser scanning microscopy showed that secondary cultures

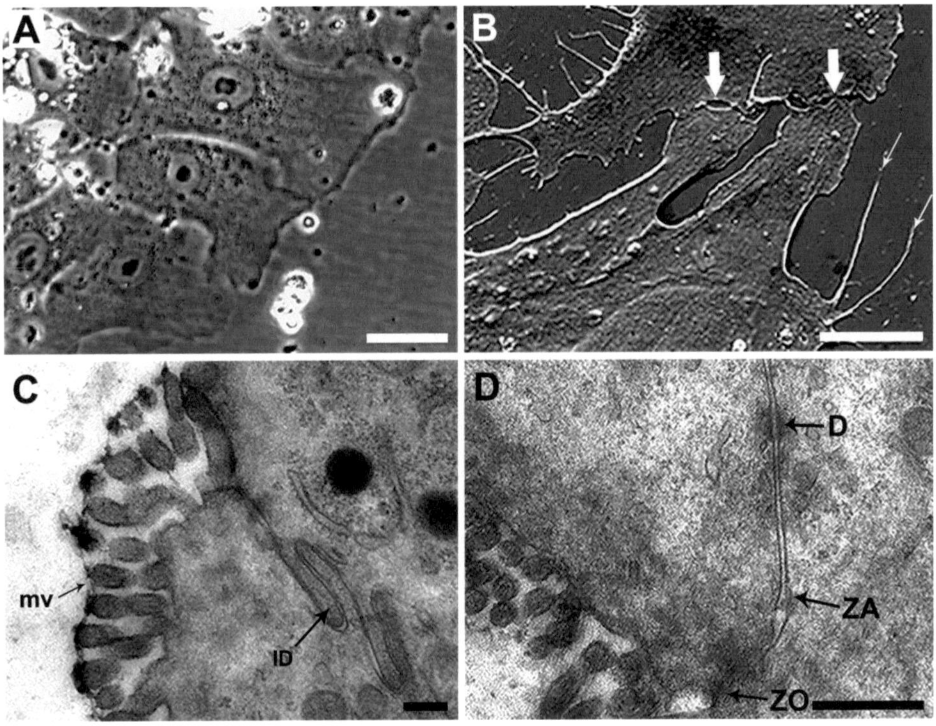

Figure 1.
Morphological characterization of cells isolated from felid small intestines. (A) Cells with 24 h of cultivation showed epithelial characteristics, such as alignment and polarization. The nuclei of the cells are located in the same plane, similar to the organization of the intestinal epithelium. (B) Scanning electron microscopy revealed thin and long cytoplasmic projections and focal points of adhesion to the substrate (thin arrows). Cell-cell contact areas form specialized contact points, such as intercellular junctions (thick arrows). (C-D) By transmission electron microscopy, many microvilli (mv) are observed at the apical pole. Junctional complex areas (ZO = zonula occludens; ZA = zonula adherens; D = desmosomes) are located at interdigitations. Bars: (A) 20 μm; (B) 5 μm; (C) 0.5 μm; (D) 0.2 μm.

of FIECs preserved the morphological and functional characteristics of immature enterocytes. These cells sustained strong expression of cytokeratin concentrated around the nuclei after two weeks, indicating they were truly epithelial (**Figure 2A–D**). Double staining by phalloidin-FITC to identify actin filaments and the anti-cytokeratin antibody revealed little to no co-localization between these proteins (**Figure 2A**). The localization of actin filaments was mostly observed at focal adhesion points for the substrate at the cellular membrane (**Figure 2A–G**). The functional properties of FIECs were evaluated based on the expression of intestinal alkaline phosphatase, which is an enzyme secreted by the intestinal epithelium (**Figure 2B–D**). Intestinal alkaline phosphatase expression was initially detected after 5 days of culture (**Figure 2B**). The labeling showed a progressive increase in the enzyme concentration inside the cells, which occurred between 7 and 9 days post-cultivation (**Figure 2C–D**). The immunocytochemistry assays targeting vimentin and desmin failed in the FIECs until up to 15 days in secondary culture (data not shown), as expected for healthy intestinal cells.

3.3 *T. Gondii* bradyzoite-FIEC interaction *in vitro*

Previously, we described the behavior of bradyzoites during their interaction with FIECs [8]. Here, quantitative and qualitative analyses were performed with ratios

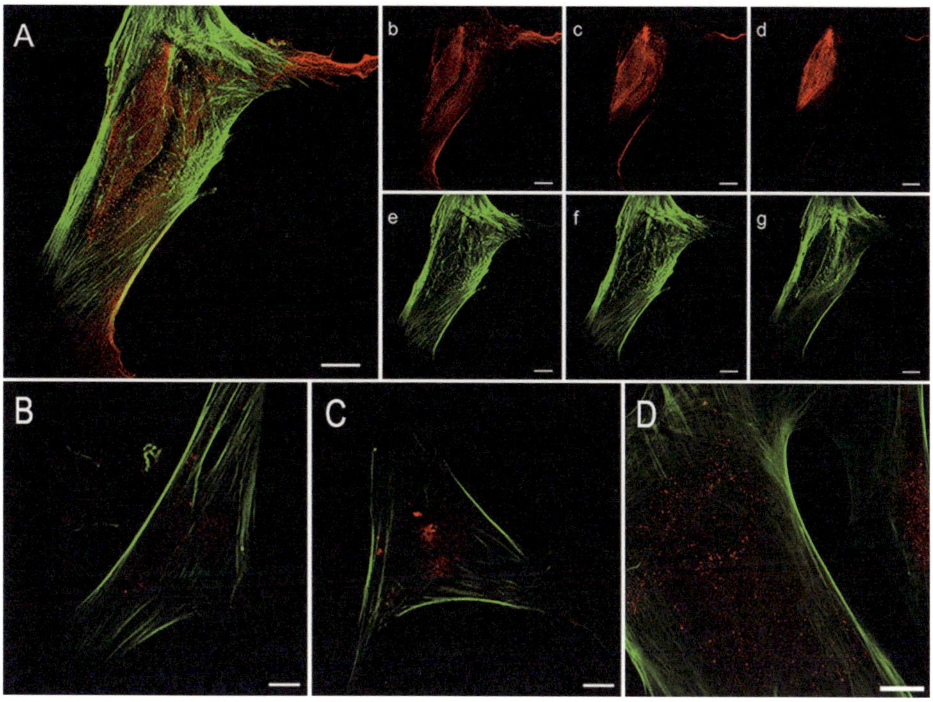

Figure 2.
Characterization of FIEC by fluorescence microscopy. (A) Culture of FIECs presenting epithelial morphology, revealed by double labeling: Actin filaments in green and cytokeratin in red. (b–d) Cytokeratin expression is located around the nucleus. (e–g) Actin filament organization revealed by phalloidin in FIECs shows the cell morphology and the major concentration at the focal adhesion points. (B–D) Intestinal alkaline phosphatase immunoreactivity in FIECs (red), shows activity that progressively increased in the cells as a function of the culture time. (B) 5 days, (C) 7 days, and (D) 9 days post-seeding. Actin filaments are shown in green. Bars: 20 µm.

of 1:5, 1:10, and 1:20 (parasite: host cell). The number of infected cells was analyzed after 24 to 96 h of parasite and host cell interaction (**Figure 3**). The data indicated the influence of the parasite load on the number of infected enterocytes during the study period: 9% after 24 hours of interaction and 42.4% after 96 hours when the 1:5 ratio was used (**Table 1**). Ratios of 1:10 and 1:20 (parasite: host cell) resulted in a lower number of infected enterocytes when compared to the 1:5 ratio. The main difference between the 1:10 and 1:20 ratios was the occurrence of structures similar to cysts (cyst-like) or schizonts (schizont-like), as previously observed by our group [8]. Cyst-like structures were more common when the ratio of 1:10 was employed, while the ratio of 1:20 resulted in more schizonts-like structures (**Table 1**).

The analysis of the parasite-host cell interaction with the 1:5 ratio revealed that the parasites doubled during the first 24 h of infection, with rosette form indicating the occurrence of endodyogeny, as seen by Giemsa (**Figure 4A**) and immunofluorescence, revealing tachyzoites with anti-SAG antibodies (**Figure 4B**). The bradyzoite-tachyzoite conversion occurred as shown by staining with the anti-SAG1-TRITC antibody (**Figure 4B**). After 96 hours of interaction, parasites were found in the extracellular environment, characterizing the lytic cycle of *T. gondii*, as seen by Giemsa (**Figure 4C**) and immunofluorescence (**Figure 4D**).

The establishment of *T. gondii* cystogenesis with the ratio of 1:10 was confirmed (**Figure 5**). Cyst-like intracellular structures in enterocytes were visible after 48 h

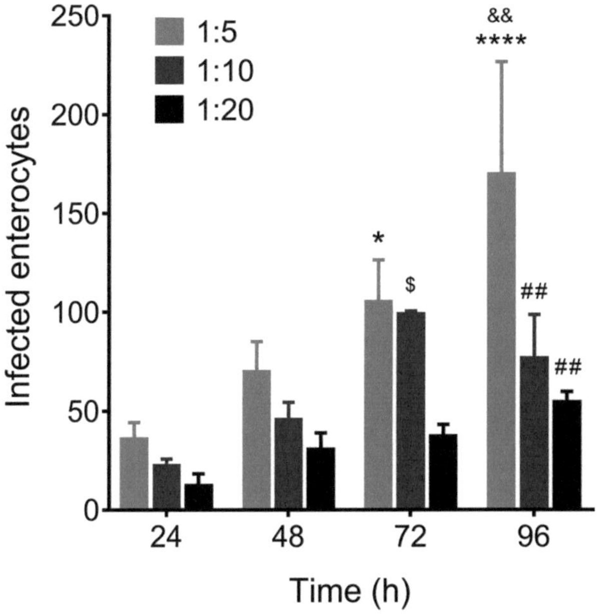

Figure 3.
Absolute number of FIECs infected with T. gondii ME-49 strain bradyzoites. A total number of 400 cells were counted for each coverslip. The parasite-host cell ratios (MOI) were 1:5, 1:10 and 1:20. **** *p < 0.0001 and* * *p < 0.05 compared to 24 h at MOI 1:5.* $^{\&\&}$ *p < 0.01 compared to 48 h at MOI 1:5.* $^{\$}$ *p < 0.05 compared to 24 h at MOI 1:10.* $^{\#\#}$ *p < 0.01 compared to MOI 1:5 at 96 h.*

Hours of interaction	Parasite:host cell ratio (MOI)	% Infected cells (IC)		% Cyst-like (CL)		% Schizont-like (SL)	
		Mean	SD	Mean	SD	Mean	SD
24	1:5	9.0	3.6	0	0	0	0
	1:10	5.6	1.5	0	0	0	0
	1:20	3.1	2.6	0	0	0	0
48	1:5	17.5	6.6	1.4	2.4	0	0
	1:10	11.4	3.8	11.7	10.2	0.8	1.4
	1:20	7.7	3.6	11.2	15.6	5.4	1.9
72	1:5	26.3	9.2	2.6	1.1	0	0
	1:10	24.8	0.7	32.1	16.0	0.7	1.2
	1:20	9.3	2.6	24.4	21.0	12.9	6.1
96	1:5	42.4	24.6	1.9	1.0	0	0
	1:10	19.2	9.5	75.3	76.2	0	0
	1:20	13.7	2.2	17.0	26.9	7.6	3.4

Table 1.
Quantitative analyses of cyst-like and schizont-like forms during FIEC infection with ME49 T. gondii bradyzoites. Time comparison of IC for MOI 1:5: 96 h x 24 h *****, 96 h x 48 h* ***, 72 h x 24 h* **. Time comparison of IC for MOI 1:10: 72 h x 24 h* **. MOI comparison of IC for 96 h: 1:20 x 1:5 and 1:10 x 1:5* ***. Time comparison of SL for MOI 1:20: 72 h x 24 h* *****; 96 h x 24 h and 72 h x 48 h* ****; 96 h x 72 h and 48 h x 24 h* ***. Time comparison of CL for MOI 1:10: 96 h x 24 h* ***; 96 h x 48 h* **. MOI comparison of SL for 48 h: 1:20 x 1:10 and 1:20 x 1:5* **. MOI comparison of SL for 72 h: 1:20 x 1:10 and 1:20 x 1:5* *****. MOI comparison of SL for 96 h: 1:20 x 1:10 and 1:20 x 1:5* ****. MOI comparison of CL for 96 h: 1:20 x 1:10* **; 1:10 x 1:5* ***.* $^{*}p < 0.05$, $^{**}p < 0.01$, $^{***}p < 0.001$, $^{****}p < 0.0001$.

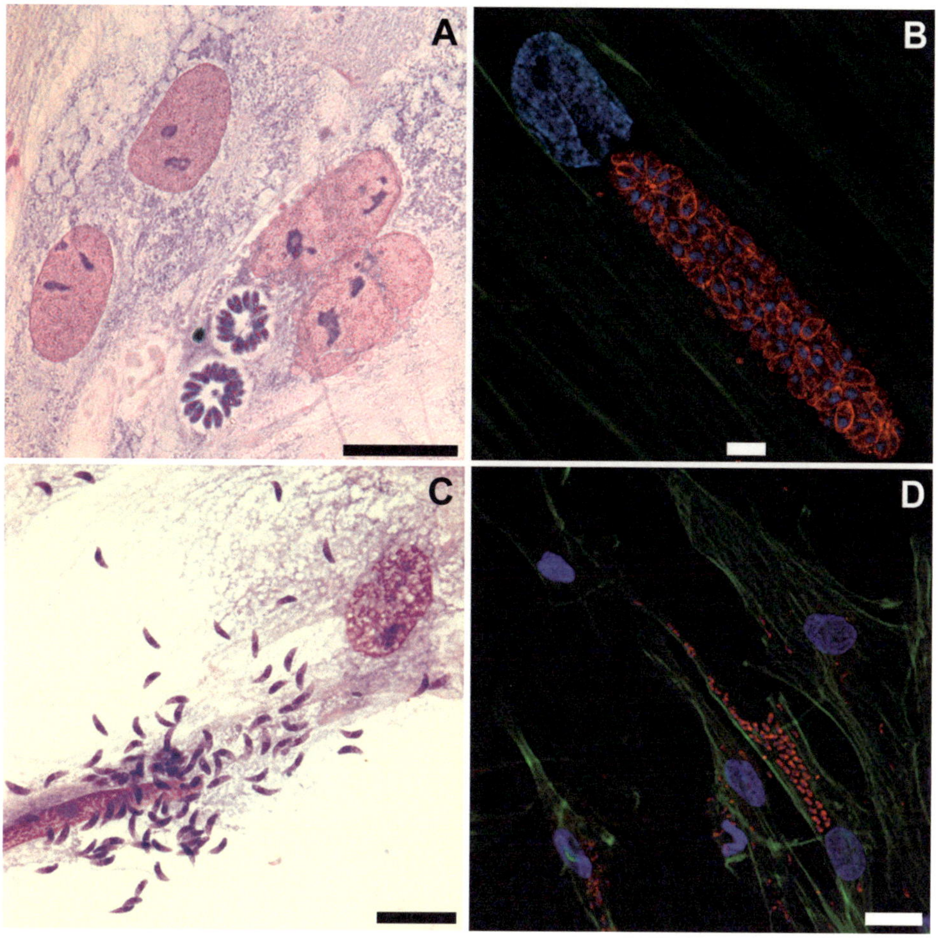

Figure 4.
Light microscopy of feline enterocytes infected with bradyzoites of T. gondii (1:5 parasite:host cell ratio). (A) Parasitophorous vacuoles show parasites in classic rosettes, indicating the interconversion from bradyzoites to tachyzoites. (B) Immunostaining for SAG1 (in red) reveals tachyzoites in FIECs and actin filaments (in green) with phalloidin. (D, E) The establishment of the lytic cycle was observed at 96 h provoked by tachyzoite proliferation based on Giemsa staining and immunolabeling with anti-SAG1 antibody. Bars: 20 µm.

(**Table 1**) by differential interference contrast microscopy (**Figure 5A**), staining with Giemsa (**Figure 5B**), and confirmed by ultrastructural analysis showing cyst wall, bradyzoites containing various granules of amylopectin and electron-dense rhoptries (**Figure 5C**).

As described during the quantitative analyses, schizont-like forms of *T. gondii* were observed in FIECs with the 1:20 ratio (**Table 1**). Starting at 48 h, cultures stained with Giemsa presented large parasitophorous vacuoles containing multinucleated masses (**Figure 6**). These structures had wide variability of shape and size, resembling the schizonts found in the gut of felines during the process of schizogony. It was common to observe more than one of these multinucleated masses (**Figure 6A, E**) and cell division processes by endodyogeny and schizogony in the same cell (**Figure 6D, E**). These structures are characterized as C-type schizonts, as advocated by Speer & Dubey [25], in a process of multiple nuclear division, with migration from the nuclei to the periphery, as can be seen in **Figure 6A–D**. This process of schizogony gives rise to merozoites,

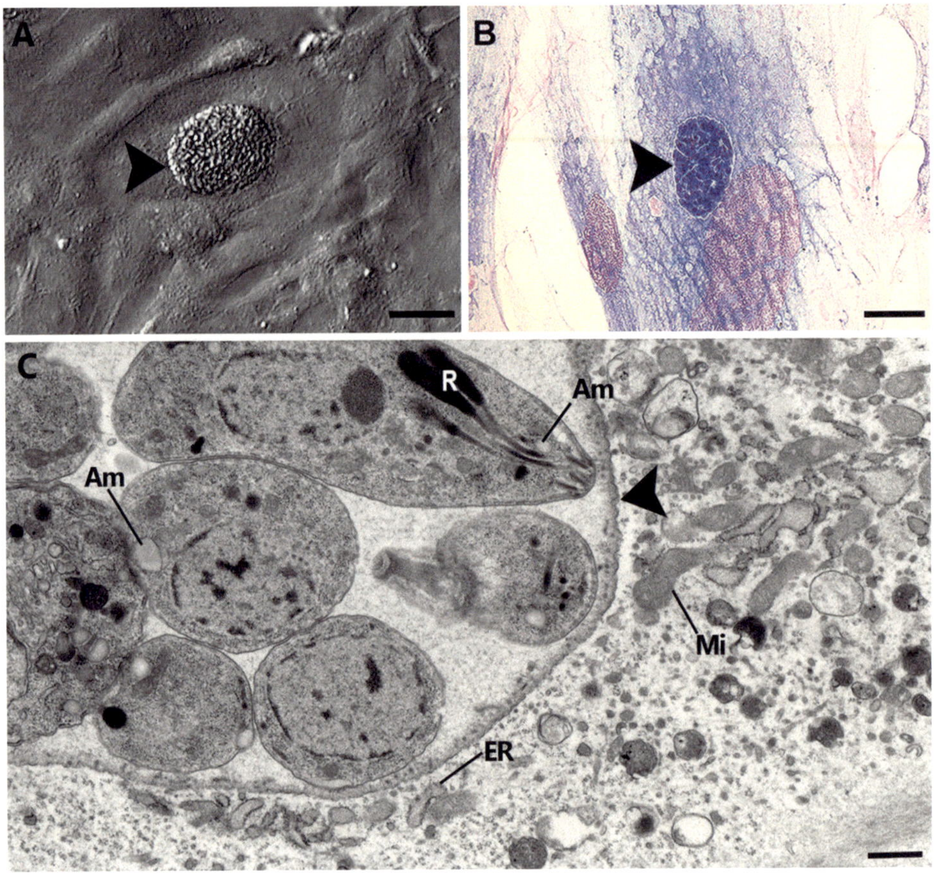

Figure 5.
Cysts of T. gondii in FIECs infected with the 1:10 bradyzoites:host cell ratio. (A-B) Differential interference contrast (DIC) microscopy (A) and Giemsa staining (B) show cyst-like structures (arrowheads) after 72 hpi (C) Cyst of T. gondii revealed by TEM with the cyst wall (arrowhead), parasites containing amylopectin granules (Am), and electron-dense rhoptries (R). ER = endoplasmic reticulum; Mi = mitochondria. Bar: (a, B) 20 μm; (C) 0.5 μm.

which are organized immediately below the schizont's membrane, as in a "budding," as suggested by **Figure 6F**. These will be better presented during the ultrastructural analyses. With the advancement of schizogony, the merozoites acquired a peripheral arrangement (**Figure 7A–D**) with the presence (**Figure 7B–D**) or not of a residual body (**Figure 7A**). The occurrence of different pathways of intracellular fate of the parasite was observed in a single cell: lytic cycle with formation of rosettes (**Figure 7B, D**) simultaneously to vacuoles containing C-type schizonts (**Figure 7A–D**).

Another type of schizont detected in enterocytes *in vitro,* according to the description of *in vivo* models, was a group of three or four parasites in a parallel arrangement, with or without residual body, identified as D-type schizonts. These structures were seen in PV isolated in the cytoplasm of cells, along with other PV-containing C-type schizonts of *T. gondii* (**Figure 8**).

Ultrastructural analysis of these infected cultures for periods ranging from 48 h to 9 days showed PV containing parasites with morphological characteristics similar to those of *T. gondii* enteroepithelial stages. Evidence of schizonts containing merozoites supports the suggestion that partial reproduction of the enteric cycle of *T. gondii* was

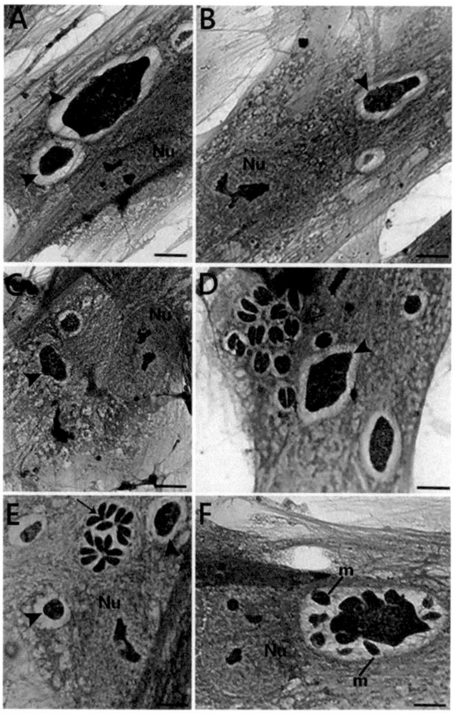

Figure 6.
Enterocytes infected with T. gondii bradyzoites of the ME-49 strain for 48 h. It is possible to observe several multinucleated masses of varied shapes and sizes in large vacuoles, corresponding to type C schizonts (arrowheads). (D, E) The same cell presents parasites in pairs (arrow in D) or rosettes (arrow in E) in process of cell division, corresponding to the lytic cycle, and vacuoles containing multinucleated masses (arrowheads). (F) Type C schizont showing merozoites (m) emerging from the multinucleated mass. Nu = enterocyte nucleus. Bars: 10 μm.

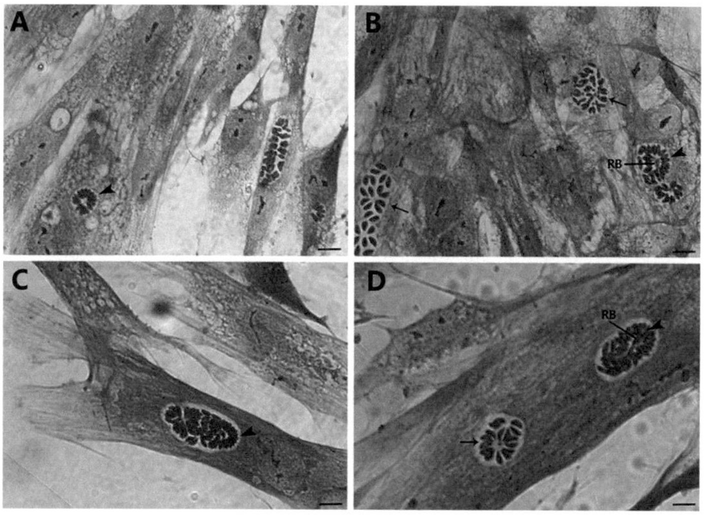

Figure 7.
Enterocytes infected with T. gondii bradyzoites of the strain ME-49 for 72 h. Note the presence of parasitophorous vacuoles that are morphologically smaller, along with peripherally arranged parasites inside the vacuoles with or without a large residual body (RB), as described for type C schizonts (arrowhead). Other vacuoles containing parasites in different stages of development can be seen in neighboring cells or in the same cell (arrows). Bars: 10 μm.

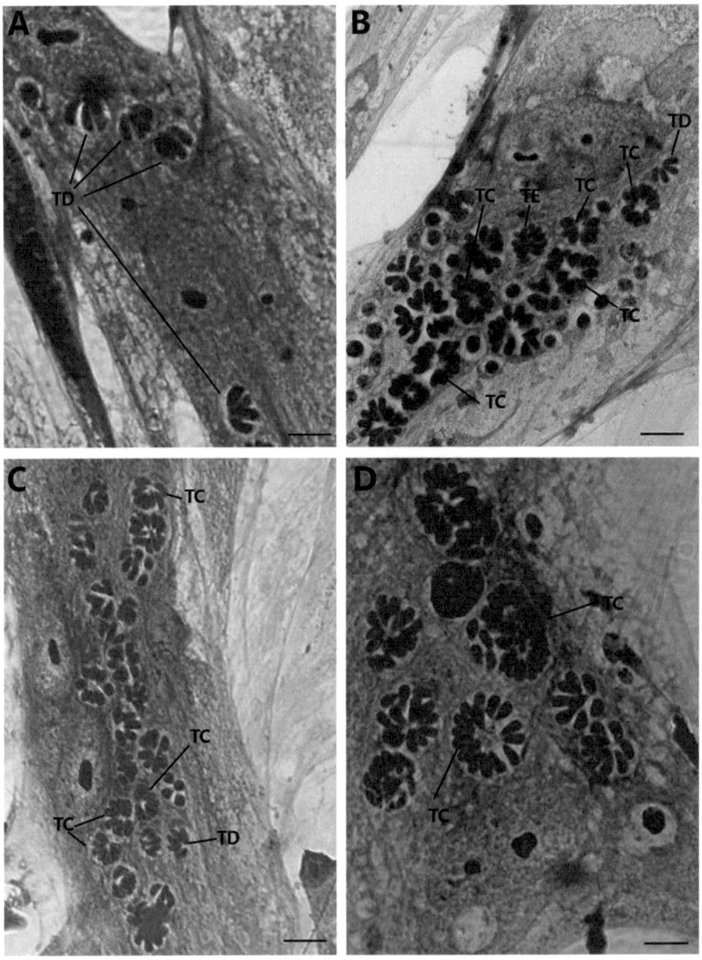

Figure 8.
Enterocytes infected with T. gondii bradyzoites of the strain ME-49 for 96 h. (A–C) Images suggestive of type D schizonts (TD) in groups of 3–4 parasites with or without residual bodies. (B–D) Several vacuoles containing parasites in different stages of cell division and development. Type C schizonts (TC) show parasites arranged peripherally inside the parasitophorous vacuoles. Bars: 10 μm.

obtained *in vitro*, as will be detailed below. Electron micrographs showed enterocytes with multinucleated structures that correspond to the C-type schizont forms (**Figure 9**), similar to those detected here by light microscopy stained with Giemsa (**Figure 6**).

The ultrastructural morphological characteristics of these multinucleated masses showed a varying number of nuclei in each of these structures, with diverse sizes and shapes, higher incidence of rounded shapes, presence of voluminous dense granules, and lipid bodies (**Figure 9**). These forms also presented a well-developed tubulovesicular membrane network (TMN) in the vacuolar matrix, best seen in **Figure 9A**, **C**, and **E**. From these multinucleated masses in a certain stage of development of the endopolygeny or schizogony, the merozoites constructed within these masses began to migrate to the periphery, like "budding" on the surface of these structures, being easily recognized by the presence of the emerging conoid, and these stages were identified as type C schizonts (**Figure 9**). **Figure 10** is suggestive of the formation of D-type schizonts, as ascertained by Giemsa-stained light microscopy (**Figure 8A–C**).

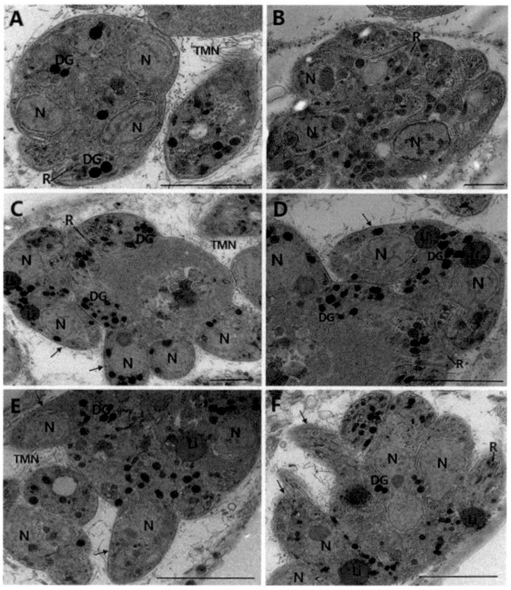

Figure 9.
Enterocytes infected with T. gondii bradyzoites of the strain ME-49 for 96 h. Ultrastructure of enterocytes containing type C schizonts. Large vacuoles containing multinucleate mass with the presence of dense granules (DG) and lipids (Li). Tubulovesicular membrane network (TMN) is present in the vacuolar space. Merozoites are seen emerging from the multinucleated masses (arrows). N = nucleus; R = rhoptries. Bars: (A, D, and E) 3 μm; (B and C) 1 μm; (F) 2 μm.

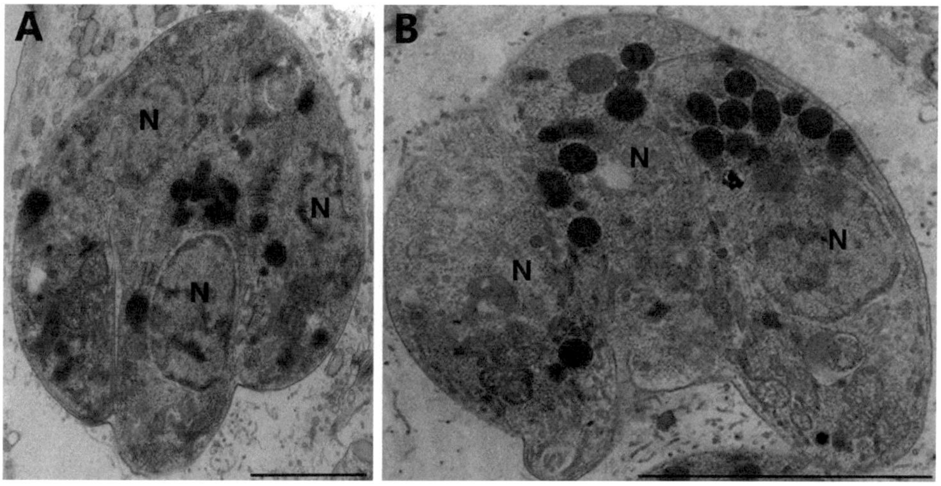

Figure 10.
Enterocytes infected with T. gondii bradyzoites from strain ME-49 for 144 h. Ultrastructure of D-type schizont-containing enterocytes. N = nucleus. Bars: (A) 2 μm; (B) 5 μm.

4. Discussion

The universal distribution of toxoplasmosis and the important role of felines in the transmission of *T. gondii* stimulate research to understand its enteroepithelial cycle better. In this context, the present study employed primary cultures of feline intestinal epithelial cells as a model to investigate the *T. gondii*-host cell interaction.

Some methodological aspects employed in the present study deserve special attention. The use of bradyzoites as a source of infection is justified because it represents one of the natural routes of transmission of *T. gondii* (through the consumption of raw meat by carnivorous animals or humans), based on the hypothesis that the parasite's enteric cycle is most efficient when cats consume tissue cysts (carnivores) [32]. Ferguson, through schematic diagrams, showed all possible routes of development and parasite conversion stages that could occur among the various forms of infection during the life cycle of *T. gondii* [24]. He demonstrated that the only infectious stage capable of direct conversion into merozoites was bradyzoites. These data support the choice of bradyzoites as a source of infection in the present study, making possible the successful differentiation of bradyzoites into merozoites *in vitro*, evidenced by light microscopy and transmission electron microscopy.

In vitro experiments have shown that some parasites initially replicate quickly as tachyzoites to amplify the infection, independent of the infective form (tachyzoites, bradyzoites, or sporozoites) [33]. The conversion of bradyzoites into tachyzoites is a natural process that occurs beginning at 15 h in cell cultures without the addition of immunomodulatory substances [34]. We confirmed this process in FIECs infected with ME49 strain bradyzoites at the 1:5 (parasite: host cell) ratio and, to a smaller extent, 1:10 and 1:20 ratios. It was possible to analyze the intracellular fate of *T. gondii* in the three routes: lytic cycle, cystogenesis, and schizogony in these ratios for periods ranging from 3 to 9 days of interaction.

Our results revealed that decreasing the parasite ratio to 1:10 (bradyzoite: host cell) caused the spontaneous formation of well-defined intracellular cysts in enterocytes after 72 h without any modulation (physical, chemical, or immunological) of the cell culture. Like other researchers, we consider that cystogenesis is a spontaneous event dependent on the strain of *T. gondii*. For example, low-virulent strains (type II) such as ME49 have a natural ability to form cysts in mammalian cells [35–40]. Therefore, we believe that the cell type might be one of the factors that determine the intracellular parasite fate and that intrinsic cellular factors could promote the differentiation stage of *T. gondii* without the need for extrinsic stress factors [41–43].

Here, the occurrence of cystogenesis in feline enterocytes was well characterized ultrastructurally. Our group has already demonstrated in epithelial cells that infection of the feline renal epithelial line CRFK with bradyzoites of strain ME49 (the same strain used in the present study) was more efficient in the establishment of cystogenesis compared to the mouse intestinal epithelial cell line IEC-6 [44]. Our data, combined with the fact that FIEC differentiates in culture, reinforces the concept that cystogenesis *in vitro* is contingent upon several factors, including the strain, parasite load, and cell type (or the interaction of these factors), in addition to the stage of differentiation of the host cells [44–47].

The experimental conditions applied in our experiments using bradyzoites of a low-virulent strain of *T. gondii*, such as ME49, and feline enterocytes allowed us to obtain infective stages corresponding to the morphological characteristics of schizont forms of the parasite, very similar to those characterized *in vivo* [24, 25]. The reproducibility of the sexual cell cycle of *Isospora suis* in intestinal swine epithelial cells was obtained before [48], and it was pointed out the significant influence of the infective dose on the development of intracellular merozoites. The researchers obtained a high density of merozoites when the 1:10 ratio (parasite: host cell) was used and it even allowed the production of oocysts *in vitro* with low parasite loads (1:100 or 1:200). These data are corroborated in part by the results of our group [8] that FIEC infection with bradyzoites of the ME49 strain using different parasite:host cell ratios was decisive for the intracellular fate of the parasites in enterocytes, in particular, to obtain schizonts. Several experiments were carried out to induce higher production of schizonts in feline enterocytes to, perhaps, obtain sexual forms of *T. gondii*. In this study, we used parasitic load variation, but we were faced with the difficulty of observing a sufficient number of infected cells to allow ultrastructural analysis. We, therefore, chose to infect with the 1:20 load to explore this interaction better. Another strategy also employed was the reduction in the concentration of fetal bovine serum to 1% in the culture medium, as suggested before for *Isospora* [48], but in our system, there was no apparent influence on further induction of schizogony.

Our analysis by optical microscopy of Giemsa-stained monolayers was elucidating because the images of the intracellular stages had a close morphological correlation with the description of schizont stages traditionally reported in the gut of experimentally infected cats [17, 24–26, 49]. The development of schizonts and gametogony in feline enterocytes has been established from the infection of cats with bradyzoites, giving rise to the various stages of *T. gondii* schizonts [49]. It has been postulated that bradyzoites, when penetrating enterocytes, trigger the production of five distinct enteroepithelial stages or schizonts, which are conventionally called types A, B, C, D, and E [25].

The presence of large numbers of multinucleated masses in our enterocyte cultures was the first indication that the *T. gondii* enteroepithelial cycle was established *in vitro*. These structures would correspond to C-type schizonts in the multiple multiplication process, characterized as endopolygeny and/or schizogony by optical microscopic analysis and confirmed by observing the ultrastructure, as proposed by Speer and colleagues [25, 28, 50]. These C-type schizonts are characterized by TMN, mitochondria, electron-dense rhoptries, a large volume of lipids, amylopectin granules, and nuclei displaced to the periphery, as shown in **Figure 9**, in agreement with Speer and Dubey's descriptions [25]. The comparative analysis of the intracellular structures evidenced in enterocytes *in vitro* (**Figure 6A** and **9A**) revealed high similarity with those described by Ferguson [26], who characterized the enteroepithelial stages of histological sections of the intestines of cats. Later stages of the developing schizont type C in enterocyte cultures (**Figure** 7) are similar to those seen in sections of intestinal tissue from infected cats [25], confirming that we were able to partly reproduce the enteroepithelial cycle of *T. gondii in vitro*. Representative images of the schizogony in the process of "budding" of merozoites were observed by electron microscopy (**Figure 9E, F**). When compared to the corresponding ones from this type of schizont described in the original article by Ferguson [26], these images indicate the morphological similarities between these structures. Light and electron microscopic analysis enabled the identification of D-type schizonts, which are also characterized by multinucleated masses that give rise to merozoites from asymmetric nuclear division, which generates organisms with various morphological aspects [17]. Comparing the enterocyte culture images of **Figure 8A** and **10A** with those of histological sections of infected cat intestine [25] shows similarities, indicating that D-type schizonts were produced *in vitro*.

Thus, under our experimental conditions, cultures of FIECs infected with bradyzoites revealed structures very similar to the schizonts of types C and D according to the classification established by Dubey and Frenkel [17] and Speer and Dubey [25] in histological sections of the small intestine of cats orally infected with *T. gondii* cysts.

5. Conclusion

The experimental strategies implemented in this work reproduced *in vitro* the natural cellular microenvironment necessary to establish the enteric development of *T. gondii* in the definitive host, the domestic cat. The primary feline intestinal epithelial cell culture indicated the potential contribution to new approaches for the investigation of parasite cell biology. We demonstrated that analysis of FIECs is an alternative method that could be used to understand the enteric cycle of *T. gondii* under controlled conditions, thereby opening the field for an investigation into the molecular aspects of this interaction. This approach could contribute to the development of new strategies aimed at intervention targeting one of the main routes by which toxoplasmosis spreads.

Acknowledgements

We thank Sandra Maria de Oliveira Souza, Rômulo Custodio dos Santos, and Genesio Lopes de Faria for technical assistance, and also Pedro Paulo Manso and Marcelo Pelajo from Confocal Platform of PDTIS-Fiocruz for help with the Zeiss LSM 510 META confocal laser scanning microscope.

This work was supported by grants from Conselho Nacional de Desenvolvimento Científico e Tecnológico (CNPq), Fundação Carlos Chagas Filho de Amparo à Pesquisa do Estado do Rio de Janeiro (FAPERJ), Fundação Oswaldo Cruz (Programa Estratégico de Apoio à Pesquisa em Saúde - PAPES VI), Programa de Apoio a Núcleos de Excelência (Pronex) CNPq/FAPERJ, Edital FAPERJ N° 15/2019 Apoio a Redes de Pesquisa em Saúde no Estado do Rio de Janeiro, and Instituto Oswaldo Cruz/Fiocruz. English review and revision by Mhali Translators.

Author details

Renata M. de Muno[1†], Marcos A. Moura[1,2†], Letícia C. Medeiros[1], Pedro N. Caldas[3], Rafael M. Mariante[1] and Helene S. Barbosa[1*]

1 Laboratory of Structural Biology, Oswaldo Cruz Institute, Rio de Janeiro, Brazil

2 School of Medical Sciences and Health of Juiz de Fora, Juiz de Fora, MG, Brazil

3 Veterinary Hospital Niteroi, Niterói, RJ, Brazil

*Address all correspondence to: helene@ioc.fiocruz.br

† These authors contributed equally to this work.

IntechOpen

© 2022 The Author(s). Licensee IntechOpen. This chapter is distributed under the terms of the Creative Commons Attribution License (http://creativecommons.org/licenses/by/3.0), which permits unrestricted use, distribution, and reproduction in any medium, provided the original work is properly cited.

References

[1] Mahida YR, Galvin AM, Gray T, Makh S, McAlindon ME, Sewell HF, et al. Migration of human intestinal lamina propria lymphocytes, macrophages and eosinophils following the loss of surface epithelial cells. Clinical and Experimental Immunology. 1997;**109**:377-386. DOI: 10.1046/j.1365-2249.1997.4481346.x

[2] Rusu D, Loret S, Peulen O, Mainil J, Dandrifosse G. Immunochemical, biomolecular and biochemical characterization of bovine epithelial intestinal primocultures. BMC Cell Biology. 2005;**6**:42. DOI: 10.1186/1471-2121-6-42

[3] Macartney KK, Baumgart DC, Carding SR, Brubaker JO, Offit PA. Primary murine small intestinal epithelial cells, maintained in long-term culture, are susceptible to rotavirus infection. Journal of Virology. 2000;**74**:5597-5603. DOI: 10.1128/jvi.74.12.5597-5603.2000

[4] Aldhous MC, Shmakov AN, Bode J, Ghosh S. Characterization of conditions for the primary culture of human small intestinal epithelial cells. Clinical and Experimental Immunology. 2001;**125**:32-40. DOI: 10.1046/j.1365-2249.2001.01522.x

[5] Quaroni A. Fetal characteristics of small intestinal crypt cells. Proceedings of the National Academy of Sciences of the United States of America. 1986;**83**:1723-1727. DOI: 10.1073/pnas.83.6.1723

[6] Perreault N, Beaulieu JF. Use of the dissociating enzyme thermolysin to generate viable human normal intestinal epithelial cell cultures. Experimental Cell Research. 1996;**224**:354-364. DOI: 10.1006/excr.1996.0145

[7] Sanderson IR, Ezzell RM, Kedinger M, Erlanger M, Xu ZX, Pringault E, et al. Human fetal enterocytes *in vitro*: Modulation of the phenotype by extracellular matrix. Proceedings of the National Academy of Sciences of the United States of America. 1996;**93**:7717-7722. DOI: 10.1073/pnas.93.15.7717

[8] Moura Mde A, Amendoeira MR, Barbosa HS. Primary culture of intestinal epithelial cells as a potential model for *Toxoplasma gondii* enteric cycle studies. Memórias do Instituto Oswaldo Cruz. 2009;**104**:862-864. DOI: 10.1590/s0074-02762009000600007

[9] Simon-Assmann P, Turck N, Sidhoum-Jenny M, Gradwohl G, Kedinger M. *In vitro* models of intestinal epithelial cell differentiation. Cell Biology and Toxicology. 2007;**23**:241-256. DOI: 10.1007/s10565-006-0175-0

[10] Desmarets LM, Theuns S, Olyslaegers DA, Dedeurwaerder A, Vermeulen BL, Roukaerts ID, et al. Establishment of feline intestinal epithelial cell cultures for the propagation and study of feline enteric coronaviruses. Veterinary Research. 2013;**44**:71. DOI: 10.1186/1297-9716-44-71

[11] Evans GS, Flint N, Potten CS. Primary cultures for studies of cell regulation and physiology in intestinal epithelium. Annual Review of Physiology. 1994;**56**:399-417. DOI: 10.1146/annurev.ph.56.030194.002151

[12] Tenter AM. *Toxoplasma gondii* in animals used for human consumption. Memórias do Instituto Oswaldo Cruz. 2009;**104**:364-369. DOI: 10.1590/s0074-02762009000200033

[13] Schlüter D, Däubener W, Schares G, Groß U, Pleyer U, Lüder C. Animals

are key to human toxoplasmosis. International Journal of Medical Microbiology. 2014;**304**:917-929. DOI: 10.1016/j.ijmm.2014.09.002

[14] Kravetz J. Congenital toxoplasmosis. BMJ. Clinical Evidence. 2013;**2013**:0906

[15] Halonen SK, Weiss LM. Toxoplasmosis. Handbook of Clinical Neurology. 2013;**114**:125-145. DOI: 10.1016/B978-0-444-53490-3.00008-X

[16] Dubey JP, Navarro IT, Sreekumar C, Dahl E, Freire RL, Kawabata HH, et al. *Toxoplasma gondii* infections in cats from Paraná, Brazil: Seroprevalence, tissue distribution, and biologic and genetic characterization of isolates. The Journal of Parasitology. 2004;**90**:721-726. DOI: 10.1645/GE-382R

[17] Dubey JP, Frenkel JK. Cyst-induced toxoplasmosis in cats. The Journal of Protozoology. 1972;**19**:155-177. DOI: 10.1111/j.1550-7408.1972.tb03431.x

[18] Dubey JP. Feline toxoplasmosis and coccidiosis: A survey of domiciled and stray cats. Journal of the American Veterinary Medical Association. 1973;**162**:873-877

[19] Dubey JP, Frenkel JK. Experimental toxoplasma infection in mice with strains producing oocysts. The Journal of Parasitology. 1973;**59**:505-512

[20] Frenkel JK, Dubey JP, Miller NL. *Toxoplasma gondii* in cats: Fecal stages identified as coccidian oocysts. Science. 1970;**167**:893-896. DOI: 10.1126/science.167.3919.893

[21] Ferguson DJ, Hutchison WM, Dunachie JF, Siim JC. Ultrastructural study of early stages of asexual multiplication and microgametogony of *Toxoplasma gondii* in the small intestine of the cat. Acta Pathologica et Microbiologica Scandinavica. Section B: Microbiology and Immunology. 1974;**82**:167-181. DOI: 10.1111/j.1699-0463.1974.tb02309.x

[22] Ferguson DJ, Hutchison WM, Siim JC. The ultrastructural development of the macrogamete and formation of the oocyst wall of *Toxoplasma gondii*. Acta Pathologica et Microbiologica Scandinavica. Section B. 1975;**83**:491-505. DOI: 10.1111/j.1699-0463.1975.tb00130.x

[23] Koyama T, Shimada S, Ohsawa T, Omata Y, Xuan X, Inoue N, et al. Antigens expressed in feline enteroepithelial-stages parasites of *Toxoplasma gondii*. The Journal of Veterinary Medical Science. 2000;**62**:1089-1092. DOI: 10.1292/jvms.62.1089

[24] Ferguson DJ. Use of molecular and ultrastructural markers to evaluate stage conversion of *Toxoplasma gondii* in both the intermediate and definitive host. International Journal for Parasitology. 2004;**34**:347-360. DOI: 10.1016/j.ijpara.2003.11.024

[25] Speer CA, Dubey JP. Ultrastructural differentiation of *Toxoplasma gondii* schizonts (types B to E) and gamonts in the intestines of cats fed bradyzoites. International Journal for Parasitology. 2005;**35**:193-206. DOI: 10.1016/j.ijpara.2004.11.005

[26] Ferguson DJ. *Toxoplasma gondii*: 1908-2008, homage to Nicolle, Manceaux and Splendore. Memórias do Instituto Oswaldo Cruz. 2009;**104**:133-148. DOI: 10.1590/s0074-02762009000200003

[27] Mehran M, Levy E, Bendayan M, Seidman E. Lipid, apolipoprotein, and lipoprotein synthesis and secretion during cellular differentiation in Caco-2 cells. In Vitro Cellular & Developmental

Biology. Animal. 1997;**33**:118-128. DOI: 10.1007/s11626-997-0032-3

[28] Freyre A. Separation of toxoplasma cysts from brain tissue and liberation of viable bradyzoites. The Journal of Parasitology. 1995;**81**:1008-1010

[29] Guimarães EV, Acquarone M, de Carvalho L, Barbosa HS. Anionic sites on *Toxoplasma gondii* tissue cyst wall: Expression, uptake and characterization. Micron. 2007;**38**:651-658. DOI: 10.1016/j.micron.2006.09.002

[30] Popiel I, Gold MC, Booth KS. Quantification of *Toxoplasma gondii* bradyzoites. The Journal of Parasitology. 1996;**82**:330-332

[31] Bohne W, Gross U, Ferguson DJ, Heesemann J. Cloning and characterization of a bradyzoite-specifically expressed gene (hsp30/bag1) of *Toxoplasma gondii*, related to genes encoding small heat-shock proteins of plants. Molecular Microbiology. 1995;**16**:1221-1230. DOI: 10.1111/j.1365-2958.1995.tb02344.x

[32] Dubey JP. Comparative infectivity of oocysts and bradyzoites of *Toxoplasma gondii* for intermediate (mice) and definitive (cats) hosts. Veterinary Parasitology. 2006;**140**:69-75. DOI: 10.1016/j.vetpar.2006.03.018

[33] Dubey JP. Survival of *Toxoplasma gondii* tissue cysts in 0.85-6% NaCl solutions at 4-20°C. The Journal of Parasitology. 1997;**83**:946-949

[34] Gross U, Bohne W, Lüder CG, Lugert R, Seeber F, Dittrich C, et al. Regulation of developmental differentiation in the protozoan parasite *Toxoplasma gondii*. The Journal of Eukaryotic Microbiology. 1996;**43**:114S-116S. DOI: 10.1111/j.1550-7408.1996.tb05033.x

[35] McHugh TD, Gbewonyo A, Johnson JD, Holliman RE, Butcher PD. Development of an *in vitro* model of *Toxoplasma gondii* cyst formation. FEMS Microbiology Letters. 1993;**114**:325-332. DOI: 10.1111/j.1574-6968.1993.tb06593.x

[36] Dardé ML, Bouteille B, Leboutet MJ, Loubet A, Pestre-Alexandre M. *Toxoplasma gondii*: étude ultrastructurale des formations kystiques observées en culture de fibroblastes humains [*Toxoplasma gondii*: Ultrastructural study of cystic formations observed in human fibroblast culture]. Annales de Parasitologie Humaine et Comparée. 1989;**64**:403-411. French. DOI: 10.1051/parasite/1989646403

[37] Lindsay DS, Dubey JP, Blagburn BL, Toivio-Kinnucan M. Examination of tissue cyst formation by *Toxoplasma gondii* in cell cultures using bradyzoites, tachyzoites, and sporozoites. The Journal of Parasitology. 1991;**77**:126-132

[38] Guimarães EV, de Carvalho L, Barbosa HS. Primary culture of skeletal muscle cells as a model for studies of *Toxoplasma gondii* cystogenesis. The Journal of Parasitology. 2008;**94**:72-83. DOI: 10.1645/GE-1273.1

[39] Guimarães EV, Carvalho L, Barbosa HS. Interaction and cystogenesis of *Toxoplasma gondii* within skeletal muscle cells *in vitro*. Memórias do Instituto Oswaldo Cruz. 2009;**104**:170-174. DOI: 10.1590/s0074-02762009000200007

[40] de Muno RM, Moura MA, de Carvalho LC, Seabra SH, Barbosa HS. Spontaneous cystogenesis of *Toxoplasma gondii* in feline epithelial cells *in vitro*. Folia Parasitol (Praha). 2014;**61**:113-119

[41] da Silva F, Mda F, Barbosa HS, Gross U, Lüder CG. Stress-related

and spontaneous stage differentiation of *Toxoplasma gondii*. Molecular BioSystems. 2008;**4**:824-834. DOI: 10.1039/b800520f

[42] Ferreira-da-Silva Mda F, Rodrigues RM, Andrade EF, Carvalho L, Gross U, Lüder CG, et al. Spontaneous stage differentiation of mouse-virulent *Toxoplasma gondii* RH parasites in skeletal muscle cells: An ultrastructural evaluation. Memórias do Instituto Oswaldo Cruz. 2009;**104**:196-200. DOI: 10.1590/s0074-02762009000200012

[43] Ferreira-da-Silva Mda F, Takács AC, Barbosa HS, Gross U, Lüder CG. Primary skeletal muscle cells trigger spontaneous *Toxoplasma gondii* tachyzoite-to-bradyzoite conversion at higher rates than fibroblasts. International Journal of Medical Microbiology. 2009;**299**:381-388. DOI: 10.1016/j.ijmm.2008.10.002

[44] Molestina RE, El-Guendy N, Sinai AP. Infection with *Toxoplasma gondii* results in dysregulation of the host cell cycle. Cellular Microbiology. 2008;**10**:1153-1165. DOI: 10.1111/j.1462-5822.2008.01117.x

[45] Lavine MD, Arrizabalaga G. Induction of mitotic S-phase of host and neighboring cells by *Toxoplasma gondii* enhances parasite invasion. Molecular and Biochemical Parasitology. 2009;**164**:95-99. DOI: 10.1016/j.molbiopara.2008.11.014

[46] Kim MJ, Jung BK, Cho J, Song H, Pyo KH, Lee JM, et al. Exosomes secreted by *Toxoplasma gondii*-infected L6 cells: Their effects on host cell proliferation and cell cycle changes. The Korean Journal of Parasitology. 2016;**54**:147-154. DOI: 10.3347/kjp.2016.54.2.147

[47] Worliczek HL, Ruttkowski B, Schwarz L, Witter K, Tschulenk W, Joachim A. *Isospora suis* in an epithelial cell culture system - an *in vitro* model for sexual development in coccidia. PLoS One. 2013;**8**:e69797. DOI: 10.1371/journal.pone.0069797

[48] Dubey JP, Miller NL, Frenkel JK. *Toxoplasma gondii* life cycle in cats. Journal of the American Veterinary Medical Association. 1970;**157**:1767-1770

[49] Speer CA, Dubey JP, Blixt JA, Prokop K. Time lapse video microscopy and ultrastructure of penetrating sporozoites, types 1 and 2 parasitophorous vacuoles, and the transformation of sporozoites to tachyzoites of the VEG strain of *Toxoplasma gondii*. The Journal of Parasitology. 1997;**83**:565-574

[50] Speer CA, Clark S, Dubey JP. Ultrastructure of the oocysts, sporocysts, and sporozoites of *Toxoplasma gondii*. The Journal of Parasitology. 1998;**84**:505-512

Chapter 3

Understanding the Effects of Toxoplasmosis on Host Behavior, Personality, and Cognition

Ruth Adekunle and Almeera Lateef

Abstract

Toxoplasma gondii is a parasite that affects about 20–80% of the global population. Chronic infection with toxoplasma, also called latent infection, has largely been considered to be asymptomatic with minimal to no clinical effects or sequelae. Though there is now clear evidence in animal models and mounting evidence in humans that latent toxoplasmosis can have various effects on behavior, personality, cognition, and even psychiatric conditions. In this chapter, we will explore the role latent toxoplasmosis plays in the behavior of animals and humans, and discuss the possible mechanisms for the observed effects.

Keywords: toxoplasmosis, animal behavior, human behavior, personality, cognition, mechanism, neuropsychiatric conditions

1. Introduction

Since its discovery in 1908, there has been curiosity about the interplay between toxoplasmosis, behavioral changes, personality changes, and cognitive changes in both animals and humans. Toxoplasmosis is one of the most common parasites globally. After primary infection with toxoplasmosis, it lives in a latent form usually in the nervous system and muscle tissues of the host. Animal studies demonstrate that latent infection with toxoplasmosis can have effects on the behavior and overall performance of animals. Human studies also suggest that latent toxoplasmosis infection affects personality, behavior, and cognition, and likely plays a role in the development of psychiatric conditions, particularly schizophrenia. It is hypothesized that modification of the host behavior helps to promote transmission of the parasite. Through multiple intricate experiments, the mechanisms of how latent infection affects behavior and psychomotor performance are felt to be largely driven by altered levels of dopamine. It has been also been postulated that humans with certain blood groups who are infected with *Toxoplasma* are three times more likely to be in a traffic accident while others are protected from the effects of toxoplasmosis infection. The ubiquitous nature of toxoplasma and easily accessible methods to determine infection, allows it to be an ideal parasite to study not only the manipulation theory in animal models but to better understand the subtle ways toxoplasmosis may also be affecting human behavior.

2. Effect of toxoplasma on different hosts

2.1 Cats as the definitive hosts of toxoplasma

Toxoplasma has an indirect life cycle. This means that cats are the only host where *Toxoplasma gondii* can complete its reproduction cycle in which *T. gondii* undergoes full gametogenesis and mating in the feline intestine. This results in the generation of oocytes that are shed by cats [1, 2]. Several species of rodents, small birds, and warm-blooded animals including, humans, dogs, rabbits, mice, rats, wild birds, and sea otters, for example, can ingest the oocytes. When this occurs, the parasite undergoes asexual reproduction and the oocytes can contaminate several different environments. The distribution of oocytes in the environment can lead to further infection by other hosts, with hopes of eventually returning to the feline species. Since the reproduction life cycle of *T. gondii* can only be completed in cats, T. gondii has developed strong selective pressures to evolve mechanisms that enhance transmission from the intermediate host to the definitive cat host. These mechanisms involve altering animal behavior, activity, personality, and cognition. The modification of the host's behavior is known as the "manipulation hypothesis" [2].

2.2 Effect of toxoplasma on infected mice and rats

Early studies looking at behavioral changes in rodents noted a decreased learning capacity and impaired memory. Piekarski et al. performed maze experiments with rats and mice who were infected with *Toxoplasma.* The authors noted that the learning performance of rats and especially mice were impacted and that the memory of mice and their overall activity was decreased. These were some of the initial observations that highlighted the influence of parasitic infections on [2]the behavior of hosts [3].

Cats are immediately attracted to moving and exposed objects, thus it would behoove *T. gondii* to increase host activity and decrease fear of open spaces. Hutchison et al. performed a series of tests to investigate this hypothesis. The authors looked at three groups of mice infected with *Toxoplasma*: one group was infected and the other two groups were infected congenitally. In the two congenitally infected groups, one was infected before mating and the other was infected during gestation. These mice were then compared with uninfected mice. Each mouse was tested in a box, the floor of which was marked off into 16 equal squares, and its activity was measured over 10 minutes by counting the number of times the mouse entered each square. Behavior was categorized as moving, rearing, digging, grooming and immobility. The results demonstrated that infection with *Toxoplasma* was associated with an increase in the amount of general movement but decreased rearing and digging. This demonstrated that the effect *T. gondii* has on behavior is selective. In addition to increased activity, specifically an increase in short bouts of behavior type, infected mice showed a relative preference for being in the more exposed or new areas of the apparatus boxes [4–6]. These behaviors were not felt to be fully explained by behavioral abnormalities such as lowered motivation or general debility as it would be unlikely for these traits to consistently produce increased levels of activity. It is theorized that *Toxoplasma*-infected mice interact with their environment and the stimulation arising from it in a different way than uninfected mice. Furthermore, this study highlighted that the mode of infection, whether congenital or latent did not differ from the behavioral changes noted in infected mice as all infected mice, irrespective of the mode of infection, showed increased activity compared to uninfected controls [5].

Webster et al. then performed a series of experiments assessing the effects of *Toxoplasma* on rats. Rats are unlike mice in that they exhibit neophobic behavior or the avoidance of new stimuli. Webster et al. used four groups of rats: two groups were placed in cages where the rat's reaction to three food-related novel stimuli (odor, food container, and food) was performed. The other two groups underwent a trappability study to evaluate whether *Toxoplasma* affected the probability of capture. The authors noted that the mice demonstrated low neophobia and that low neophobia was significantly associated with positive *Toxoplasma* titer in three out of four groups of rates. Additionally, *Toxoplasma*-infected rats were more susceptible to trapping and poisoning during post-control programs [7].

Webster et al. then went on to test the correlation and relationship between parasite transmission and load with rat behavioral changes. He sought to understand two things: (1) whether parasites with indirect life cycles, will influence the activity of the intermediate host, increasing the likelihood of transmission and (2) whether the change in observed activity would be an increase in activity rather than a decrease. The authors used four groups of mice: one group were wild brown rats infected with naturally occurring parasites, another was a hybrid of wild and laboratory rats that were experimentally infected with *T. gondii* as adults, another was wild and laboratory rats that were experimentally infected with *T. gondii* congenitally, the last group were rats where were infected with *Syphacia muris*. The difference between *T. gondii* and *Syphacia muris* is that *T. gondii* has an indirect life cycle and *Syphacia muris* has a direct life-cycle parasite. The rats were also matched with uninfected hybrid rats. In the wild occurring rats, out of the six-parasite species detected, *T. gondii* was the only parasite to be associated with higher activity levels when comparing infected rats to uninfected rats. The hybrid rats also experienced increased activity levels compared with uninfected rats. Lastly, there were no differences in activity levels noted between the rats infected with *Syphacia muris* compared with uninfected mice. This study provided evidence that the indirect life-cycle of *T. gondii* can influence the activity of an intermediate host, such as the rat [8].

One of the most profound effects of *T. gondii* infection is the ability to minimize and or eliminate the natural aversion to the odor of predators. In the wild, rodents are at risk of being attacked and eaten by many species, including cats, foxes, and mink. Because of this, rodents have developed an evolutionally innate aversion to the odor of these predators, decreasing their risk of predation [9]. Even laboratory rodents maintain this aversion. Though this aversion to host, particularly cat urine, presents itself as an obstacle to parasites with indirect life cycles such as *T. gondii*. Experiments performed by Berdoy et al. were some of the first to demonstrate *T. gondii*'s ability to alter this aversion to cat odor and decrease the predatory risk of cats, now coined "fatal feline attraction." To test this theory, the authors performed a nocturnal exploratory test on the behavior of rats in outdoor pens. All outdoor pens were covered in a neutral homogeneous and neutral surface. In each corner of the outdoor pen, drops of four distinct types of smells were placed: the rat's urine, cat urine, rabbit urine (served as a control for a mammalian non-predator), and water (neutral smell). The authors noted a significant difference in the behavior of *Toxoplasma*-infected and *Toxoplasma*-uninfected rats where uninfected rats maintained an aversion to predatory urine, while infected rats showed a preference for areas containing cat urine. Both groups of rats behaved similarly to the smell of their urine, water, and rabbit odor. The authors concluded that this potentially fatal attraction was not a gross impairment of olfactory senses, but rather a subtle change in the cognitive perception of the host in the face of predatory risk [10]. The findings of Berdoy et al. have been replicated in other studies [11–13]. Kaushik et al. further demonstrated that this "fatal feline attraction" is not specific to just one feline species, but is exhibited by both domestic cats and wild cats. The authors also reported the interesting finding that not all feline odors were equal but *Toxoplasma*-infected rats had a stronger preference for wild cat odor over that of domestic cats. This effect did not differ significantly according to the type of wild cat odor used (cheetah or puma) [9].

2.3 Large animals

T. gondii's effect on behavior expands across a broad range of warm-blooded species. In a study performed by Gering et al., they showed that *T. gondii* can affect the behavior of hyenas (*Crocuta crocuta*) and their naturally occurring interactions with lions (a feline species). Out of the 166 surveyed hyenas, 108 had IgG antibodies to T. *gondii*. The other hyenas were determined to either be uninfected or doubtful to be infected with *Toxoplasma*. From a demographic standpoint, the only differences noted between infected and uninfected hyenas were that infection rates were lowest in cubs (35%), followed by subadults (71%) and adults had the highest rate of infection (80%). They then investigated the association of *T. gondii* infection with boldness toward lions which was determined by their minimum approach distance to the lion(s). Cubs and adult hyenas were analyzed separately because older hyenas consistently approach lions more closely than cubs. Infected cubs had a shorter minimum approach distance from lions (−3.19, 95% CI: −5.57 – 0.81]) than their uninfected counterparts, though for subadults and adults, infection was not related to minimum approach distance. Lastly, the authors assess the association *between T. gondii* infection with lion-related mortality. Among 33 mixed-age hyenas with known mortality causes, infected hyenas were nearly twice as likely to die by lions than by other known causes (52% vs. 25%). When translated into odds, infected hyenas were 3.91 times more likely to die by lions compared with uninfected hyenas, though this funding was not statistically significant (95% CI: 0.70–32.78; $P = 0.15$). In a sub-analysis performed on the 11 cubs infected with *T. gondii*, 100% of the deaths were caused by lions, while only 17% of the uninfected cub deaths were caused by lions. The authors concluded that *T. gondii* infection was associated with behavioral boldness that brought infected hyena cubs into closer proximity to lions, increasing the likelihood of being killed by lions [14].

Recently, mild red foxes have been demonstrated to show uncharacteristic behavior which has been classified as Dopey Fox Syndrome (DFS). These behaviors include an apparent lack of fear, increased affection, constant pacing, facial muscle twitching, and anorexia. Conditions such as encephalitis as well as visual abnormalities and/or blindness which are consistent with *Toxoplasma*, raise the possibility of *T. gondii* being a causative agent. Milne et al. investigated the association between *T. gondii* and/or other neurotropic agents, with DFS. Serology and PCR targeting *T. gondii* were used to determine Toxoplasma infection and a multiplex PCR was developed to test for other neutropenic agents. Some examples of the other neurotropic agents used include canine-specific circovirus, a fox-specific circovirus, canine herpes virus, canine distemper virus, and canine adenovirus type 1 and 2. Results were compared between foxes with DFS maintained in welfare facilities to foxes provided by local pest-control agencies, which the authors referred to as "pseudo-control" foxes. The authors noted that the prevalence of *T. gondii* infection was higher in captive foxes (33%) compared with pseudo-control foxes (6.6%). In the behavioral analysis, the results indicated that *Toxoplasma*-infected foxes preferred cat odor zones compared to dog odor zones.

Additionally, noted was a lack of fear of humans and increased inquisitiveness among *Toxoplasma*-infected captive foxes. The authors though did also find evidence of possible co-infection with other agents playing a role in DFS. In a small sample of captive and pseudo-control foxes that were euthanized (occurred outside of the study setting), there was a higher prevalence of *Toxoplasma* and fox-specific circovirus co-infection in captive foxes (33%) compared with pseudo-control foxes (11%). Co-infection was also associated with aberrant fox behavior, suggesting possible synergy between these two agents that contribute to the presence of DFS [15].

There is a suggestion that the effects of *T. gondii* on human dates back to when our ancestors were still under significant feline predation. Poirotte et al. evaluated whether or not chimpanzees, the human's closest relative, experienced any attraction to feline urine. They found that *Toxoplasma*-infected chimpanzees have lost their innate aversion toward the urine of leopards, who are their only natural predator. This finding seemed to be selective for *T. gondii* as there were no significant differences in the response of *Toxoplasma*-infected and *Toxoplasma*-uninfected chimpanzees toward urine collected from other definitive feline hosts that chimpanzees do not encounter in nature [16]. It is plausible these behavioral modifications could increase the probability of chimpanzee mortality by leopards for the parasite's benefit.

3. Toxoplasmosis effect on human personality and behavior

3.1 Role of toxoplasma on human behavior and personality

Toxoplasmosis is known to affect the activity of rodents, though there are also well-established studies implicating an association between Toxoplasmosis and human behavioral changes. Toxoplasmosis in humans can be in three forms: congenital, acute, or chronic. Several studies outline the devasting effects Toxoplasmosis can have on a fetus, ranging from hydrocephalus, chorioretinitis, and intracranial calcifications to fetal demise. Reduced intellection function has been reported in approximately 6–9% of children with congenital toxoplasmosis [6, 17–19]. Acute infection has been associated with psychosis confusion, aphasia, and other space-occupying neurologic symptoms [20, 21]. There has been increasing interest in the role that chronic or latent infection has on human personality and behavior.

Flegr et al. performed some of the early studies looking at the role latent toxoplasmosis plays in personality changes. One of his first studies looked at the correlation between serologic evidence of *T. gondii* and personality using the Cattell's personality test among 338 individuals. The authors reported a correlation between *Toxoplasma* and two personalities: lower scores of low superego strength (disregards rules, expedient) and higher scores of Protension (suspecting, jealous, dogmatic). This finding was particularly seen in males [22]. The authors further expanded their study by looking at 224 men and 170 women. In men, they noted similar personality shifts as the prior study, though in addition guilt proneness (apprehensive, self-reproaching, insecure) and group dependency (sociably group dependent, "joiner") were also positively influenced in *Toxoplasma*-infected men. In women, shifts in personality traits include an increase in affectothymia (warm-hearted, outgoing, easygoing), alaxia (trusting, accepting conditions, tolerant), untroubled adequacy (self-assured, placid, secure, complacent), and self-sufficiency (self-sufficient, resourceful, prefers own decisions). The authors wanted to tease out if the association between *Toxoplasma* infection and personality traits were a result of toxoplasmosis inducing the personality shift or if certain personality factors increased the likelihood of becoming infected with *T. gondii*. To understand this better, the authors examined the personality profiles of 164 male patients who were diagnosed with acute toxoplasmosis within the past 13 years. They were able to conclude that the positive correlation between the duration of latent toxoplasmosis and the intensity of superego strength decrease ($P < 0.020$) suggested that the decrease of superego strength (the willingness to accept group moral standards) was induced by *T. gondii* infection [23]. While the results of these studies are interesting and intriguing, others have criticized the selection methods of the population, the interpretation of the data, and the fact that the results have not been duplicated to date [6]. In later studies, using Cloninger's TCI personality tests, *Toxoplasma*-infected individuals demonstrated decreased scores on factor NS (novelty seeking). This means that individuals had a lower tendency to search for new stimuli, which was seen in both men and women [24–26]. Decreased novelty-seeking scores have been associated with increased concentration of dopamine in the brain tissue, which is consistent with the increased levels of dopamine since in the brain tissue of infected mice [25, 27].

Prolonged reaction time secondary to latent toxoplasmosis has been well established in rodents, though there is increasing data highlighting this same phenomenon of impaired psychomotor performance in humans. One such study was performed by Flagr et al. where increased human activity was associated with increased traffic accidents. The authors conducted a retrospective study assessing the association between toxoplasmosis infection and traffic accidents. The participants included 146 who experienced a traffic accident to 446 persons in the general population. While the seroprevalence of toxoplasmosis varied by age, latent toxoplasmosis was significantly higher in the traffic accident set ($P < 0.0001$). The authors discovered that subjects with latent toxoplasmosis had 2.65 times higher risk of a traffic accident than the toxoplasmosis-negative subjects (95% CI: 1.76–4.01). It is proposed that the prolongation of reaction times caused by toxoplasmosis increases the risk of incidents such as traffic accidents [28]. Additional behavioral adaptations seen in patients with Toxoplasmosis include impaired long-term concentration, lower tidiness in *Toxoplasma*-infected males, lower sociability scores in *Toxoplasma*-infected males, and decreased altruism [29, 30].

3.2 The role of toxoplasma on neuropsychiatric conditions

Since the 1950s, it was noted that the prevalence of toxoplasmosis among psychiatric patients, especially patients with schizophrenia, was unusually high, implicating that *Toxoplasma gondii* may play a role in the origin and progress of psychiatric diseases [31, 32]. Furthermore, it has been suggested that the effect of latent toxoplasmosis on the risk of schizophrenia is stronger than that of any schizophrenia-associated gene variant identified in genome-wide analyses [33]. It has also been shown that *Toxoplasma*-infected patients with schizophrenia have more severe symptoms of hallucinations and delusions than *Toxoplasma*-free schizophrenia patients and there are structural differences in the brain between *Toxoplasma*-infected schizophrenia patients and *Toxoplasma*-free schizophrenia patients [34]. *Toxoplasma*-infected patients have a reduction in gray matter volume bilaterally when compared with *Toxoplasma*-free patients [35, 36]. Other mental health conditions that have been associated with Toxoplasmosis include autism attention deficit hyperactivity disorder, obsessive–compulsive disorder, antisocial personality disorder, learning disabilities, and anxiety disorders. The association between *T. gondii* and neuropsychiatric disorders will be more extensively explored in a later chapter.

3.3 Toxoplasma and neurodegenerative diseases

Toxoplasma can affect not only personality and behavior but can also contribute to neurocognitive dysfunction. Guenter et al. investigated the cognitive performance of infected young adults using a set of neuropsychological tests. The authors observed trends toward reduced cognitive functions but the differences did not amount to statistical significance [37]. Neurocognitive dysfunction though may be more prominent in older adults. Gajewski et al. conducted a double-blinded neuropsychological study on seniors with asymptomatic latent infection. The authors compared 42 individuals aged greater than 65 with a positive anti-*Toxoplasma* IgG to 42 individuals of the same age range that were negative for toxoplasmosis. Using a computer-based working-memory test (2-back) and several standardized psychometric tests of memory and executive cognitive functions, they determined that *Toxoplasma*-positive seniors showed an impairment of different aspects of memory. In *Toxoplasma*-positive seniors, working memory was decreased by about 35% ($P = 0.020$), they had a lower performance in a verbal memory test, both regarding immediate recall (10% reduction; $P = 0.022$), delayed recognition (6%; $P = 0.037$) and recall from long-term memory assessed by the word fluency tests (12%; $P = 0.029$). Executive functions though were not different among the two groups [38]. Hann et al. confirmed similar findings in their meta-analysis. The meta-analysis included 13 studies, comprising a population of 13,289 healthy participants with a mean age of 47. *Toxoplasma*-negative individuals performed more favorably in four cognitive domains: processing speed ($P = 0.001$), working memory ($P = 0.002$), short-term verbal memory ($P < 0.001$), and executive functioning ($P = 0.030$) [39]. Findings of this study suggest that *Toxoplasma* seropositivity is associated with mild cognitive impairment in several cognitive domains.

These findings led to further investigation into the effect *Toxoplasma* has on neurocognitive disorders, in particular, Alzheimer's disease (AD). AD results from a reduced amount of beta-amyloid plaque deposition and irreversible loss of neurons in the brain. The clinical consequence is gradual loss of memory, increasing impairment of language and other cognitive functions, and later stages with motor dysfunction and inability to perform activities of daily living [40]. Early studies did not demonstrate an association between Toxoplasmosis infection and AD. Mahami-Oskouei et al. performed a case–control study that included 75 patients with AD and 75 negative control patients and they aimed to assess if there was any correlation between toxoplasmosis infection and AD. Toxoplasmosis infection was defined as evidence of a positive anti-*Toxoplasma* IgG. Among the participants, 61.3% of Alzheimer's patients and 62.6% of healthy volunteers were positive for anti-*Toxoplasma* IgG, but all participants were negative for anti-*Toxoplasma* IgM. Their analysis did not demonstrate any significant differences between Alzheimer's patients with their controls in terms of anti-*Toxoplasma* IgG antibody ($P = 0.5$) [41]. In the same year, Perry et al. noted similar findings. Their study included 105 subjects with AD and 114 controls. Anti-*Toxoplasma* IgG antibodies were present in 41% of the AD group and 33% in the control group. They too found no association with either the prevalence of anti-*Toxoplasma* IgG antibodies among the two groups ($P = 0.25$) or the log-transformed antibody concentration ($P = 0.85$) [42].

However, two recent meta-analyses demonstrated an association between *Toxoplasma* and AD. Chegeni et al. meta-analysis investigating the possible association between *T. gondii* and AD included 8 articles. They conducted a random effects model to determine the odds ratio (OR). The results showed a common OR of 1.53 (95% CI: 1.07–2.18). They concluded that *T. gondii* can be considered a risk factor for the development of AD and exacerbation of its symptoms. It was acknowledged that three studies mainly drove the results of the meta-analysis, thus with the small number of published relevant studies being low, the risk of the presence of publication bias is relatively high [43]. Bayani et al. looked at the association between *Toxoplasma*, AD, and Parkinson's Disease. Parkinson's Disease (PD) is a neurologic condition that affects movement, often causing stiffness, tremors, and difficulty with balance and coordination. They included 11 studies (7 for PD and 4 for AD). The meta-analysis showed no statistically significant association between the presence of anti-*Toxoplasma* IgG antibody and increased risk of Parkinson's disease (OR, 1.14; 95% CI: 0.78–1.68), though the OR for the association of the presence of anti-*Toxoplasma* IgG antibody with Alzheimer's patients, compared to control group, was (OR, 1.38; 95% CI: 0.99–1.92). The authors concluded that there was a marginally significant association between *Toxoplasma* infection and Alzheimer's disease [44].

The varying results found in studies could in part be explained by *Toxoplasma* having differential effects on cognition throughout one's lifespan. Given that neurodegeneration is generally seen at an older age, on average elderly seropositive adults have been infected for much longer than younger individuals. This allows for the indirect effects of latent toxoplasmosis to further exacerbate inflammation and neurodegeneration [45]. The mechanisms of direct and indirect effects of *Toxoplasma* will be further examined in the following section.

4. Mechanisms of toxoplasma affect human behavior

4.1 Mechanism of action for cognitive effects of toxoplasma

Dopamine is a neurotransmitter that is directly involved in the regulation of cognitive processes. It has been demonstrated that *T. gondii* increases the concentration of dopamine in the brain of infected hosts, including humans, through increasing dopaminergic signaling during both acute and latent infection [24, 45]. For these reasons, it is assumed that dopamine drives many of the associated neurocognitive observations in individuals with toxoplasmosis. Two main mechanisms seem to be at play that affects cognitive function: increased catecholamine synthesis leading to increased dopamine release (these are referred to as direct effects) and chronic inflammation leading to impaired dopaminergic neurotransmission and neurodegeneration (these are referred to as indirect effects).

Dopamine is produced in two steps from its precursor tyrosine. In the first step, tyrosine hydroxylase (TH) converts tyrosine into L-DOPA. In the second step, DOPA decarboxylase converts L-DOPA into dopamine. Any increase in the conversion of tyrosine to L-DOPA will result in increased dopamine synthesis. This is important because Gaskell et al. discovered that the genome of *T. gondii* contains two genes encoding tyrosine hydroxylase and this protozoan enzyme encompasses catalytic properties similar to those of the TH found in mammalian cells (Gaskell et al., 2009) [46]. TH is produced by *T. gondii* during the formation of the bradyzoites of the cyst stages of the life cycle. This increased production of TH is selectively specific to transmitters derived from tyrosine/L-DOPA leaving those that are not unaffected [45].

Further evidence that dopaminergic signaling is indeed the most likely cause of *Toxoplasma*-associated alterations of behavior and cognition is that dopaminergic antagonists like haloperidol seem to normalize *Toxoplasma*-induced behavioral changes in infected rodents [11, 47]. Skallova et al. examined if the behavioral response of *Toxoplasma*-infected male and female mice would be attenuated when given a selective dopamine uptake inhibitor, GBR 12909. Both genders of *Toxoplasma*-infected mice had decreased locomotor activity and decreased locomotion in the open field. Infected females displayed an increased level of exploration in the hole board test. GBR 12909 induced suppression in hole board-exploration in the infected males mitigating the effects of toxoplasmosis, but had an opposite effect on the controls [47]. Webster et al. performed a similar study, using haloperidol and/or valproic acid (an anti-psychotic and mood stabilizer, respectively. These medications are used in the treatment of mental illnesses, including schizophrenia. The authors discovered that anti-psychotic drugs were as effective as anti-T. gondii drugs in preventing behavioral changes secondary to toxoplasmosis [11].

Indirect mechanisms of action of *T. gondii* involve the use of inflammatory markers that modulate neurotransmission and associated cognitive processes. To keep *T. gondii* in the latent phase of toxoplasmosis, the immune system enhances the production of proinflammatory cytokines. The release of these substances not only prevents the dissemination of toxoplasmosis, but also impacts the levels, turnover, and efficiency of many neurotransmitters including dopamine, glutamate, and serotonin. This chronic overactivation of the immune system in response to latent infection compounded with the general effects of aging likely progresses and accelerates neurodegeneration [45]. A summary of the possible direct and indirect mechanisms and neurophysiological changes induced by chronic *T. gondii* is shown (**Figure 1**—originally published by Tedford et al).

4.2 Role of rhesus blood group

The effect of toxoplasmosis on personality and performance may also be in part secondary to a specific Rh blood type. Several studies that have been performed on pregnant women and military personnel have shown that RhD blood group positivity, especially in RhD heterozygotes, protects against various effects of latent toxoplasmosis [48–51]. Novotná et al. and Flegr et al. both noted in their respective studies that Rh-positive subjects, and RhD-positive heterozygotes, in particular, were protected against latent toxoplasmosis-induced impairment of reaction times [48, 50]. Flegr et al. also noted in a prospective study that included nearly 4000 military drivers, that Rh-negative *Toxoplasma*-infected subjects had about three times higher probability of a traffic accident than Rh-negative *Toxoplasma*-free individuals or Rh-positive individuals [49]. This finding was noted regardless of whether the Rh-positive subjects were *Toxoplasma*-free or *Toxoplasma*-infected. Lastly, Kankova et al. detected that RhD-positivity might protect infected pregnant women from excessive gestational weight gain [51].

Flegr et al. further assess the association between toxoplasmosis and the personality of RhD-negative and RhD-positive subjects. The study included 502 male soldiers of Czech nationality who underwent several tests including the N-70 questionnaire, the NEO-PI-R questionnaire, the Wiener Matrizen-Test (WMT) test of intelligence, and the OTIS test intelligence. The authors noted that *Toxoplasma*-infected subjects

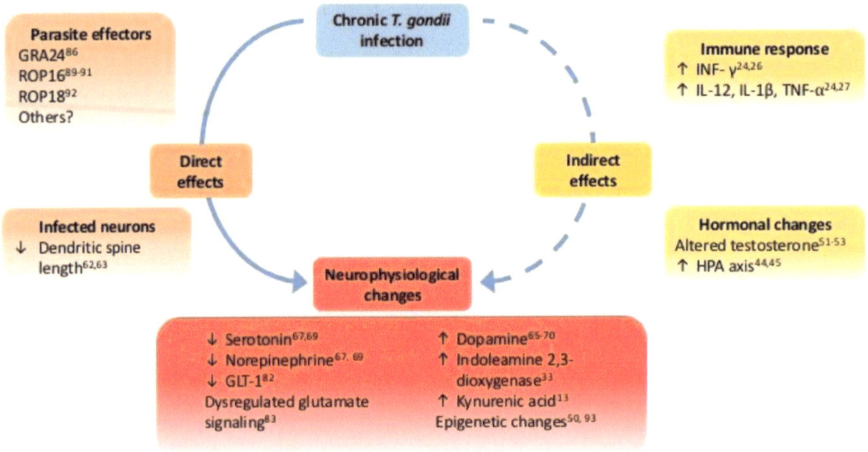

Figure 1.
Directly and indirectly mediated effects of chronic T. gondii infection on host neurophysiology. Model of mechanisms involved with host responses to infection (i.e., neuroimmune and hormonal changes) indirect and more likely confounding factors, augmenting neurophysiological changes rather than inducing them. Indeed, the specificity of behavioral changes associated with infection suggest that direct mechanisms of the parasite–host interaction play a significant role in the neurophysiological changes associated with chronic T. gondii infection.

scored lower in the total N-70 score and also in anxiety, depression, phobia, hysteria, and vegetative lability, and in the NEO-PI-R (Big Five) trait neuroticism. The differences were much stronger in RhD-negative than RhD-positive subjects. Additionally, the RhD-positive, *Toxoplasma*-infected subjects express lower verbal and nonverbal intelligence than their *Toxoplasma*-free peers. Though RhD-negative, *Toxoplasma*-infected persons expressed higher verbal and nonverbal intelligence compared with *Toxoplasma*-free peers [52].

In contrast to the findings in this study, an earlier study performed on University students noted differing results. In that study, there was no difference in expression of neuroticism between *Toxoplasma*-infected and *Toxoplasma*-free students [53]. The fact that *Toxoplasma*-infected soldiers in the Flegr et al. study expressed lower and not higher levels of psychopathognomic traits as measured by the N-70 questionnaire, was also different than in the prior study [53]. Flegr et al. went on further to determine whether the effects of RhD phenotype were exclusive to *Toxoplasma*-infected individuals or whether it influences the effects of other factors. The authors determined that the RhD phenotype modulates the influence not only of latent toxoplasmosis but also factors such as age and smoking, on human behavior and physiology. This suggests that the different human behaviors and physiology seen between RhD-positive *Toxoplasma*-infected persons and RhD-negative *Toxoplasma*-infected persons may be more related to the RhD status than the presence of *Toxoplasma* itself [54].

5. Conclusions

There is an abundant amount of data that supports Toxoplasmosis affects the behavior of animals, particularly rodents, and increasing data that supports humans also experience a variety of aberrant behaviors, personality shifts, decreased level of cognition, and development of psychiatry conditions secondary to the latent toxoplasmosis infection. However, there are a few limitations that should be considered when interpreting the data presented. It cannot be confirmed that human behavioral manipulation increases the efficiency of Toxoplasmosis transmission from intermediate to definitive hosts. Clinical trial data establishing the causality of Toxoplasmosis and behavioral modifications are lacking. It is also possible that some of the identified associations represent a trait that increases the risk of Toxoplasmosis, rather than a result of Toxoplasmosis. The extensive heterogeneity seen in the human population could influence the observed effect size in studies. Additionally, there could be synergistic effects of a third unknown agent or factor contributing to the effects of Toxoplasmosis. Despite these limitations, *T. gondii* seems to play a fascinating role in the ability to modify both animal and human behaviors with new associations between Toxoplasmosis and several conditions being frequently published. It underscores that though the effect of Toxoplasmosis infection has been studied for decades, there is still so much to be learned.

Acknowledgements

The authors' research activities are supported, in part, by grants from the National Institutes of Health/National Center for Advancing Translational Sciences (NCATS) (TL1TR002382, UL1TR002378).

Conflict of Interest

The authors declare no conflict of interest.

Author details

Ruth Adekunle* and Almeera Lateef
Department of Medicine, Division of Infectious Diseases, Medical University of South Carolina, Charleston, SC, USA

*Address all correspondence to: adekunle@musc.edu

IntechOpen

© 2022 The Author(s). Licensee IntechOpen. This chapter is distributed under the terms of the Creative Commons Attribution License (http://creativecommons.org/licenses/by/3.0), which permits unrestricted use, distribution, and reproduction in any medium, provided the original work is properly cited. (cc) BY

References

[1] Dubey JP, Cerqueira-Cézar CK, Murata FHA, Kwok OCH, Yang YR, Su C. All about toxoplasmosis in cats: The last decade. Veterinary Parasitology. 2020 Jul;**283**:109145

[2] Webster JP. The effect of *Toxoplasma gondii* on animal behavior: Playing cat and mouse. Schizophrenia Bulletin. 2007 May;**33**(3):752-756

[3] Piekarski G. Behavioral alterations caused by parasitic infection in case of latent toxoplasma infection. Zentralblatt fur Bakteriologie, Mikrobiologie, und Hygiene. Series A. 1981 Sep;**250**(3):403-406

[4] Hutchinson WM, Bradley M, Cheyne WM, Wells BW, Hay J. Behavioural abnormalities in toxoplasma-infected mice. Annals of Tropical Medicine and Parasitology. 1980 Jun;**74**(3):337-345

[5] Hay J, Hutchison WM, Aitken PP, Graham DI. The effect of congenital and adult-acquired toxoplasma infections on activity and responsiveness to novel stimulation in mice. Annals of Tropical Medicine and Parasitology. 1983 Oct;**77**(5):483-495

[6] Holliman RE. Toxoplasmosis, behaviour and personality. The Journal of Infection. 1997 Sep;**35**(2):105-110

[7] Webster JP, Brunton CF, MacDonald DW. Effect of *Toxoplasma gondii* upon neophobic behaviour in wild brown rats, *Rattus norvegicus*. Parasitology. 1994 Jul;**109**(Pt 1):37-43

[8] Webster JP. The effect of *Toxoplasma gondii* and other parasites on activity levels in wild and hybrid *Rattus norvegicus*. Parasitology. 1994 Dec;**109** (Pt 5):583-589

[9] Kaushik M, Knowles SCL, Webster JP. What makes a feline fatal in *Toxoplasma gondii*'s fatal feline attraction? Infected rats choose wild cats. Integrative and Comparative Biology. 2014 Jul;**54**(2):118-128

[10] Berdoy M, Webster JP, Macdonald DW. Fatal attraction in rats infected with *Toxoplasma gondii*. Proceedings of the Biological Sciences. 2000 Aug 7;**267**(1452):1591-1594

[11] Webster JP, Lamberton PHL, Donnelly CA, Torrey EF. Parasites as causative agents of human affective disorders? The impact of anti-psychotic, mood-stabilizer and anti-parasite medication on *Toxoplasma gondii*'s ability to alter host behaviour. Proceedings of the Biological Sciences. 2006 Apr 22;**273**(1589):1023-1030

[12] Lamberton PHL, Donnelly CA, Webster JP. Specificity of the *Toxoplasma gondii*-altered behaviour to definitive versus non-definitive host predation risk. Parasitology. 2008 Sep;**135**(10):1143-1150

[13] Vyas A, Kim SK, Giacomini N, Boothroyd JC, Sapolsky RM. Behavioral changes induced by Toxoplasma infection of rodents are highly specific to aversion of cat odors. Proceedings of the National Academy of Sciences of the United States of America. 10 Apr 2007;**104**(15):6442-6447. DOI: 10.1073/pnas.0608310104. Epub 2007 Apr 2. PMID: 17404235; PMCID: PMC1851063

[14] Gering E, Laubach ZM, Weber PSD, Soboll Hussey G, Lehmann KDS, Montgomery TM, et al. *Toxoplasma*

gondii infections are associated with costly boldness toward felids in a wild host. Nature Communications. 2021 Jun 22;**12**(1):3842

[15] Milne G, Fujimoto C, Bean T, Peters HJ, Hemmington M, Taylor C, et al. Infectious causation of abnormal host behavior: *Toxoplasma gondii* and its potential association with dopey fox syndrome. Frontiers in Psychiatry. 2020;**11**:513536

[16] Poirotte C, Kappeler PM, Ngoubangoye B, Bourgeois S, Moussodji M, Charpentier MJE. Morbid attraction to leopard urine in toxoplasma-infected chimpanzees. Current Biology. 2016 Feb 8;**26**(3):R98-R99

[17] Caiaffa WT, Chiari CA, Figueiredo AR, Orefice F, Antunes CM. Toxoplasmosis and mental retardation–report of a case-control study. Memórias do Instituto Oswaldo Cruz. 1993 Jun;**88**(2):253-261

[18] Wilson LE, Korthuis T, Fleishman JA, Conviser R, Lawrence PB, Moore RD, et al. HIV-related medical service use by rural/urban residents: A multistate perspective. AIDS Care. 2011 Aug;**23**(8):971-979

[19] Sever JL, Ellenberg JH, Ley AC, Madden DL, Fuccillo DA, Tzan NR, et al. Toxoplasmosis: Maternal and pediatric findings in 23,000 pregnancies. Pediatrics. 1988 Aug;**82**(2):181-192

[20] Monroe JM, Buckley PF, Miller BJ. Meta-analysis of anti-*Toxoplasma gondii* IgM antibodies in acute psychosis. Schizophrenia Bulletin. 2015 Jul;**41**(4):989-998

[21] Cohen BA. Neurologic manifestations of toxoplasmosis in AIDS. Seminars in Neurology. 1999;**19**(2):201-211

[22] Flegr J, Hrdý I. Influence of chronic toxoplasmosis on some human personality factors. Folia Parasitologica (Praha). 1994;**41**(2):122-126

[23] Flegr J, Zitková S, Kodym P, Frynta D. Induction of changes in human behaviour by the parasitic protozoan *Toxoplasma gondii*. Parasitology. 1996 Jul;**113**(Pt 1):49-54

[24] Flegr J, Preiss M, Klose J, Havlícek J, Vitáková M, Kodym P. Decreased level of psychobiological factor novelty seeking and lower intelligence in men latently infected with the protozoan parasite *Toxoplasma gondii* Dopamine, a missing link between schizophrenia and toxoplasmosis? Biological Psychology. 2003 Jul;**63**(3):253-268

[25] Skallová A, Novotná M, Kolbeková P, Gasová Z, Veselý V, Sechovská M, et al. Decreased level of novelty seeking in blood donors infected with toxoplasma. Neuro Endocrinology Letters. 2005 Oct;**26**(5):480-486

[26] Flegr J. Influence of latent toxoplasma infection on human personality, physiology and morphology: Pros and cons of the toxoplasma-human model in studying the manipulation hypothesis. The Journal of Experimental Biology. 2013 Jan 1;**216**(Pt 1):127-133

[27] Hodkova H, Kodym P, Flegr J. Poorer results of mice with latent toxoplasmosis in learning tests: Impaired learning processes or the novelty discrimination mechanism? Parasitology. 2007 Sep;**134**(Pt 10):1329-1337

[28] Flegr J, Havlícek J, Kodym P, Malý M, Smahel Z. Increased risk of traffic accidents in subjects with latent toxoplasmosis: A retrospective

case-control study. BMC Infectious Diseases. 2002 Jul;**2**(2):11

[29] Lindová J, Novotná M, Havlícek J, Jozífková E, Skallová A, Kolbeková P, et al. Gender differences in behavioural changes induced by latent toxoplasmosis. International Journal for Parasitology. 2006 Dec;**36**(14):1485-1492

[30] Havlícek J, Gasová ZG, Smith AP, Zvára K, Flegr J. Decrease of psychomotor performance in subjects with latent "asymptomatic" toxoplasmosis. Parasitology. 2001 May;**122**(Pt 5):515-520

[31] Kozar Z. Studies on toxoplasmosis in mental diseases. Biuletyn Państwowego Instytutu Medycyny Morskiej i Tropikalnej w Gdańsku. 1953;**5**:134-137 English transl.

[32] Torrey EF, Bartko JJ, Lun ZR, Yolken RH. Antibodies to Toxoplasma gondii in patients with schizophrenia: A meta-analysis. Schizophrenia Bulletin. May 2007;**33**(3):729-736. DOI: 10.1093/schbul/sbl050. Epub 2006 Nov 3. PMID: 17085743; PMCID: PMC2526143

[33] International Schizophrenia Consortium, Purcell SM, Wray NR, Stone JL, Visscher PM, O'Donovan MC, et al. Common polygenic variation contributes to risk of schizophrenia and bipolar disorder. Nature. 2009 Aug 6;**460**(7256):748-752

[34] Wang HL, Wang GH, Li QY, Shu C, Jiang MS, Guo Y. Prevalence of toxoplasma infection in first-episode schizophrenia and comparison between toxoplasma-seropositive and toxoplasma-seronegative schizophrenia. Acta Psychiatrica Scandinavica. 2006 Jul;**114**(1):40-48

[35] Shenton ME, Dickey CC, Frumin M, McCarley RW. A review of MRI findings in schizophrenia. Schizophrenia Research. 2001 Apr 15;**49**(1-2):1-52

[36] Horacek J, Flegr J, Tintera J, Verebova K, Spaniel F, Novak T, et al. Latent toxoplasmosis reduces gray matter density in schizophrenia but not in controls: Voxel-based-morphometry (VBM) study. The World Journal of Biological Psychiatry. 2012 Oct;**13**(7):501-509

[37] Guenter W, Bieliński M, Deptuła A, Zalas-Wiecek P, Piskunowicz M, Szwed K, et al. Does *Toxoplasma gondii* infection affect cognitive function? A case control study. Folia Parasitologica (Praha). 2012 Jun;**59**(2):93-98

[38] Gajewski PD, Falkenstein M, Hengstler JG, Golka K. *Toxoplasma gondii* impairs memory in infected seniors. Brain, Behavior, and Immunity. 2014 Feb;**36**:193-199

[39] de Haan L, Sutterland AL, Schotborgh JV, Schirmbeck F, de Haan L. Association of *Toxoplasma gondii* seropositivity with cognitive function in healthy people: A systematic review and meta-analysis. JAMA Psychiatry. 2021 Oct 1;**78**(10):1103-1112

[40] Honig LS, Mayeux R. Natural history of Alzheimer's disease. Aging (Milano). 2001 Jun;**13**(3):171-182

[41] Mahami-Oskouei M, Hamidi F, Talebi M, Farhoudi M, Taheraghdam AA, Kazemi T, et al. Toxoplasmosis and Alzheimer: Can *Toxoplasma gondii* really be introduced as a risk factor in etiology of Alzheimer? Parasitology Research. 2016 Aug;**115**(8):3169-3174

[42] Perry CE, Gale SD, Erickson L, Wilson E, Nielsen B, Kauwe J, et al. Seroprevalence and serointensity of latent *Toxoplasma gondii* in a sample of elderly adults with and without

Alzheimer disease. Alzheimer Disease and Associated Disorders. 2016 Jun;**30**(2):123-126

[43] Nayeri Chegeni T, Sarvi S, Moosazadeh M, Sharif M, Aghayan SA, Amouei A, et al. Is *Toxoplasma gondii* a potential risk factor for Alzheimer's disease? A systematic review and meta-analysis. Microbial Pathogenesis. 2019 Dec;**137**:103751

[44] Bayani M, Riahi SM, Bazrafshan N, Ray Gamble H, Rostami A. *Toxoplasma gondii* infection and risk of Parkinson and Alzheimer diseases: A systematic review and meta-analysis on observational studies. Acta Tropica. 2019 Aug;**196**:165-171

[45] Colzato L, Zhang W, Beste C, Stock AK. Dissociating direct and indirect effects: A theoretical framework of how latent toxoplasmosis affects cognitive profile across the lifespan. Neurobiology of Aging. 2021 Jun;**102**:119-128

[46] Gaskell EA, Smith JE, Pinney JW, Westhead DR, McConkey GA. A unique dual activity amino acid hydroxylase in *Toxoplasma gondii*. PLoS One. 2009 Mar 11;**4**(3):e4801

[47] Skallová A, Kodym P, Frynta D, Flegr J. The role of dopamine in toxoplasma-induced behavioural alterations in mice: An ethological and ethopharmacological study. Parasitology. 2006 Nov;**133**(Pt 5):525-535

[48] Novotná M, Havlícek J, Smith AP, Kolbeková P, Skallová A, Klose J, et al. Toxoplasma and reaction time: Role of toxoplasmosis in the origin, preservation and geographical distribution of Rh blood group polymorphism. Parasitology. 2008 Sep;**135**(11):1253-1261

[49] Flegr J, Klose J, Novotná M, Berenreitterová M, Havlíček J. Increased incidence of traffic accidents in toxoplasma-infected military drivers and protective effect RhD molecule revealed by a large-scale prospective cohort study. BMC Infectious Diseases. 2009 May;**26**(9):72

[50] Flegr J, Novotná M, Lindová J, Havlícek J. Neurophysiological effect of the Rh factor. Protective role of the RhD molecule against toxoplasma-induced impairment of reaction times in women. Neuro Endocrinology Letters. 2008 Aug;**29**(4):475-481

[51] Kaňková S, Sulc J, Flegr J. Increased pregnancy weight gain in women with latent toxoplasmosis and RhD-positivity protection against this effect. Parasitology. 2010 Oct;**137**(12):1773-1779

[52] Flegr J, Preiss M, Klose J. Toxoplasmosis-associated difference in intelligence and personality in men depends on their Rhesus blood group but not ABO blood group. PLoS One. 2013;**8**(4):e61272

[53] Lafferty KD. Can the common brain parasite, *Toxoplasma gondii*, influence human culture? Proceedings of the Biological Sciences. 2006 Nov 7;**273**(1602):2749-2755

[54] Flegr J, Geryk J, Volný J, Klose J, Cernochová D. Rhesus factor modulation of effects of smoking and age on psychomotor performance, intelligence, personality profile, and health in Czech soldiers. PLoS One. 2012;7(11):e49478

Chapter 4

Nutrient Uptake Portals in *Toxoplasma gondii* Tachyzoites

Marialice da F. Ferreira-da-Silva, Mauricio Magalhães de Paiva, Erick Vaz Guimarães and Helene S. Barbosa

Abstract

The process of nutrient acquisition by *Toxoplasma gondii* tachyzoites is an attractive target for developing and designing drugs against toxoplasmosis, however, just recently it was revealed to be an important process to be understood. The present work helps address the lack of information about the exact sites where nutrient uptake in *T. gondii*. The endocytosis of proteins by tachyzoites of *T. gondii* was measured using both fluid-phase and receptor-mediated endocytic tracers. Quantitative analysis by flow cytometry revealed important differences in the percentage of labeled parasites, incubated with BSA, dextran, or transferrin. The analysis by confocal microscopy showed that the anterior portion of the conoid is one preferential site for binding BSA and transferrin to the tachyzoite, later localized within elongated structures present in the anterior region of the parasite. The ultrastructural analysis of multiple ultrathin sections displayed the endocytic markers at the following: (i) conoid, within rhoptries, (ii) in cup-shaped invagination of the parasite membrane (micropore) and, (iii) posterior pore. The present study brings data revealing three possible nutrient uptake portals in Toxoplasma tachyzoites that may contribute in the future to a therapeutic design with a view to treatment of toxoplasmosis.

Keywords: *toxoplasma gondii*, endocytosis, nutrient uptake, tachyzoites, ultrastructural analysis

1. Introduction

Toxoplasmosis is a disease that results from the infection caused by the coccidian parasite *Toxoplasma gondii*. It is a significant public health problem worldwide. About half of the world's population is infected with Toxoplasma, but most people are asymptomatic [1]. One of the most severe manifestations of toxoplasmosis is when the acquisition of the infection occurs in the first trimester of pregnancy. The parasite can cross the placenta and reach the fetus causing congenital toxoplasmosis [2]. This infection can be systemic and result in fetal death, preterm delivery, intrauterine growth retardation, fever, pneumonia, hepatosplenomegaly, thrombocytopenia, or affect the eyes and brain [3, 4]. Considering toxoplasmosis in the current COVID-19 scenario, recently a study showed that Toxoplasma-infected patients are a greater risk of having a more severe course of the disease [5]. Despite extensive research on

IntechOpen

Toxoplasma since it's discovery in 1908, some aspects of *T. gondii* nutrient acquisition have just recently received special attention and many remain poorly understood [6]. The initial discovery on nutrient acquisition by Apicomplexa (including Toxoplasma and it's close relative Plasmodium falciparum) was from 1961, when Garnham discovered the micropore by in *P. falciparum*. Since, it has been considered a mechanism for nutrition in Apicomplexa [7]. With regard, specifically to *Toxoplasma*, only one article analyzed ultrastructurally the endocytosis of the parasite. These data have served for years as a reference, describing the role of the micropore as responsible for the incorporation of nutrients by tachyzoites and bradyzoites [8]. This structure is a continuous cup-shaped invagination of the plasma membrane located in the anterior region of the parasite. Associated with this membrane is the internal membrane complex, forming a concentric electron-dense ring that surrounds the "neck" of the micropore invagination [7]. The micropore, coated or uncoated by clathrin, is present in all three infective forms of the parasite (bradyzoites, tachyzoites, and sporozoites) [8, 9]. The rhoptries deserve to be mentioned too. They are in the number of 8–12 per cell and are the only known acidified organelles in *T. gondii* (pH of immature rhoptries is 3.5–5.5 and of mature rhoptries is 5.0–7.0). Recent studies indicate that they are most analogous to secretory lysosomal granules [10].

Despite the importance of Apicomplexa parasites, including *Toxoplasma gondii*, in public health and the fact that the nutrient acquisition process are in general attractive targets for antimicrobial drugs, knowledge of the mechanisms involved in this process in Toxoplasma is still scarce [11]. Thus, nutrient incorporation pathways and intracellular traffic in *T. gondii* are fields yet to be explored in depth. These studies could potentially contribute to a direct and specific therapy for this parasite, for the benefit of patients with disseminated toxoplasmosis.

2. Experimental design

2.1 *T. Gondii* tachyzoites isolation

Tachyzoites of *Toxoplasma gondii* (RH strain) were maintained in Swiss mice, weighing about approximately 21 g, through intraperitoneal passages with inoculum of 2 x 10^6 parasites/animal. Mice were obtained from Science and Technology for Biomodels Institute (ICTB-Fiocruz). After 48 to 72 hours of infection, the parasites were collected from the peritoneal exudate and collected in phosphate buffered saline (PBS) solution, pH 7.2. The cell suspension was centrifuged at 200 *g* for 10 min and the supernatant containing the parasites was centrifuged at 1000 *g* for 10 min and the sediment rich in *T. gondii* tachyzoites, was washed 2 or 3 times in PBS solution, pH 7.2, and quantified in a Neubauer chamber [12]. All procedures to obtain the parasites from infected mice were performed according to the Safety Standards established by the Ethical Committee for Animal Use of Fiocruz, license L-042/2018 A2.

2.2 Endocytosis assays

The endocytic capacity of *T. gondii* extracellular tachyzoites was analyzed using the following endocytic tracers:

a. fluid phase markers: (i) bovine serum albumin (BSA), a 66 kDa protein, conjugated to fluorescein (BSA-FITC) or to colloidal gold particles (BSA-Au); (ii) peroxidase, 40 kDa glycoprotein, conjugated to colloidal gold (HRP-Au) and (iii) dextran, hydrophilic polysaccharide synthesized by Leuconostoc bacteria of 4.4 kDa conjugated to TRITC (Dextran-TRITC);

b. receptor-mediated endocytic markers: transferrin, an 80 kDa protein, conjugated to fluorescein (Tf-FITC) or to colloidal gold particles (Tf-Au).

2.3 Processing for analysis by flow cytometry

Flow cytometry analysis was performed after washing the tachyzoites in PBS pH 7.2 and incubating for 15 min, 30 min, 2 hr. and 4 hr. at 37°C with 0,2 mg/ml BSA-FITC, 5 mg/ml Dextran-TRITC or 1 mg/ml Tf-FITC diluted in PBS. After three washes in PBS, the parasites were fixed for 20 min at 4°C with 4% PFA, washed 3 times for 10 min each with PBS and analyzed by flow cytometry on the same day. Non-incubated tachyzoites with the fluorochrome-labeled tracers were used to calibrate the system for morphology and granularity. Data acquisition was performed using the FACSCalibur flow cytometer (Becton & Dickinson, San Jose, USA) equipped with Cell Quest software (Joseph Trotter, Scripps Research Institute, San Diego, USA). The analyses were performed using the WinMDI2.8 program on 10,000 events acquired in a pre-established region corresponding to the parasites at the Cytometry Flux Platform at Oswaldo Cruz Institute.

2.4 Processing for analysis by laser scanning confocal microscopy

T. gondii tachyzoites freshly isolated from Swiss mice and purified from peritoneal lavage as described in 1.1 were incubated with 0,2 mg/ml BSA-FITC or Tf-FITC, for periods of 10 min, 30 min, 1 hr. or 2 hr. at 37°C, followed by washing in PBS. A drop containing the tachyzoites was incubated at 37°C for 5–15 min for parasite adhesion on a slide previously coated with poly-L-lysine. The parasites were then washed twice with PBS, followed by fixation for 5 min at room temperature with 2% paraformaldehyde (PFA). The parasites were then washed in PBS and distilled water and the coverslips mounted in DABCO (1,4 Diazabicyclo [2.2.2] octane - Triethylenediamine - "antifading"). The material was analyzed on an Olympus FV 300/BX51 laser scanning confocal microscope at the Biomanguinhos Applied Pharmacology Laboratory, Fiocruz. Fluorescence was stimulated by a 488 nm laser and 510 longpass filters combined with another 543 nm laser and 560/600 bandpass filters were applied. Differential interference contrast microscopy images were obtained simultaneously with fluorescence images in different focus planes.

2.5 Colloidal gold-protein complex

For ultrastructural analysis colloidal gold particles with an average diameter of 15 nm were obtained according to the Frens method [13]. For the formation of the colloidal gold-protein complex, the pH of the colloidal gold was adjusted to 5.5 for conjugation with albumin (BSA); 8.0 for peroxidase (HRP) and 5.0 for transferrin (Tf). The concentration of each protein required to stabilize 10 ml of colloidal gold was added and the protocol followed according to the method of Slot and Geuze (1985) [13]. Endocytic tracers were purchased by Sigma-Aldrich-St. Louis, MO, USA.

2.6 *T. gondii* tachyzoites endocytic assays by transmission electron microscopy

For ultrastructural analysis 10^8 tachyzoites were centrifuged at 1000 g for 10 min and the sediment was resuspended in PBS containing BSA-Au, HRP-Au or Tf-Au at protein/PBS ratios of 1:10. Parasites were incubated at 4°C for 20 min and then at 37°C for periods of 5 min to 4 h. After this incorporation kinetics, the solution containing tachyzoites was centrifuged at 1000 g for 10 min and the pellet washed 2 times in PBS solution. Then, tachyzoites were fixed for 30 min at 4°C at 2.5% glutaraldehyde (GA) in cacodilate buffer with 2.5 $CaCl_2$ and 3.5% sucrose, pH 7.2, washed and centrifuged three times for 15 min in the same buffer. They were post-fixed for 30 min at room temperature, with 1%, osmium tetroxide and washed with the same buffer. The parasites were dehydrated in a graded acetone series and embedded in an epoxy resin (PolyBed 812). Thin sections were stained with uranyl acetate and lead citrate and then examined under a transmission electron microscope (Jeol JEM1011) at the Rudolf Barth Electron Microscopy Platform at Oswaldo Cruz Institute.

3. Results

3.1 Bovine serum albumin

For quantitative analysis, flow cytometry was used as a tool to check the association of the fluid phase endocytosis marker, BSA-FITC to tachyzoites. The region (R1) was previously established as corresponding to the parasites (**Figure 1A**). The negative control refers to the group of parasites not incubated with the fluorochrome (**Figure 1B**). The labelling of tachyzoites with BSA-FITC was time-dependent, with percentages of 16.5%, 17.5%, 27.5%, and 32%, of labeled parasites after incubation at 37°C for 10 min, 30 min, 1 h and 2 h, respectively (**Figures 1C–F**).

The confocal microscopy analysis after incubation of *T. gondii* tachyzoites with BSA-FITC for periods of 5 min to 2 h at 37°C revealed that a small population of parasites showed labelling (**Figure 2A–D**). After 5 min incubation the marker was strictly localized in the anterior region of the parasite body, corresponding to the apical complex (**Figure 2A** and **B**). After 30 min of incubation a higher concentration of the tracer was observed in the apical region of the parasite in addition to a fine granulation with symmetrical distribution along the first third of the tachyzoite body (**Figure 2C**). Few tachyzoites showed the tracer already internalized (**Figure 2C**). Parasites kept for 2 hours at 37°C in the presence of BSA-FITC revealed the fluorescent marker located at the tip of the apical region and in a possible invagination of the body (possibly micropore) and as well intracellular marker concentrated in the posterior region corresponding to basal complex (**Figure 2D**).

Transmission electron microscopy analysis showed that BSA-Au labelling was not homogeneous among tachyzoites. Incubation for 20 min at 4°C revealed discrete labelling in some parasites with one or two gold particles associated with their membrane, particularly in the vicinity of the apical region (not shown). Parasites analyzed after incubation for 30 min at 37°C showed tachyzoites contained gold particle in a depression of the plasma membrane, showing morphological features compatible with a microspore (**Figure 3**). The intracellular localization of the BSA-Au complex was observed in rhoptries of some parasites after incubation for 1 h at 37°C (data not shown).

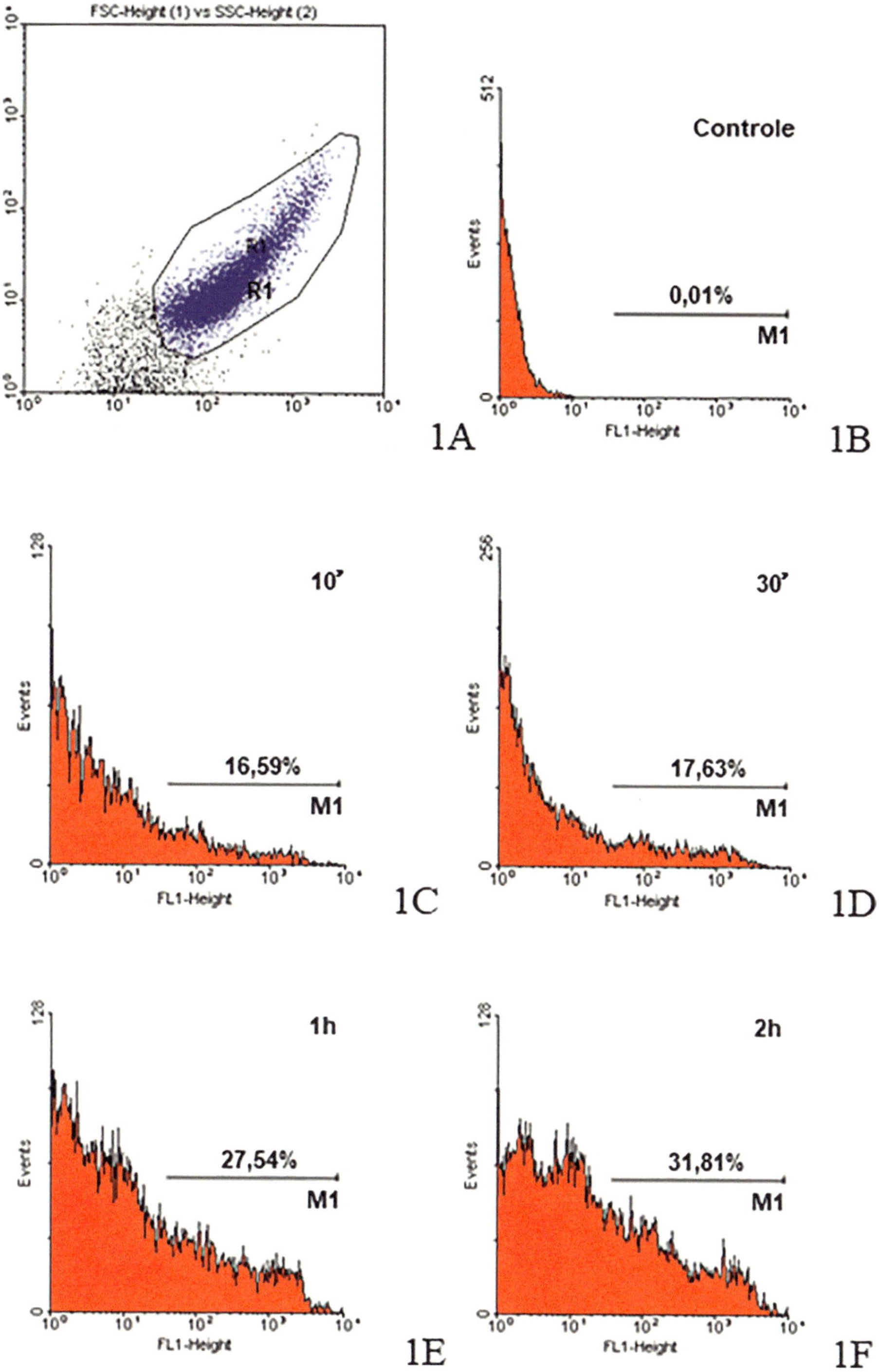

Figure 1.
Representative histograms of flow cytometry showing the kinetic of BSA-FITC internalization by T. gondii *tachyzoites incubated at 37°C for different lengths of time. (A-B) Parasites incubated with PBS. Graphics showing the morphology of the parasites. (A) Size and granularity (FSC x SSC) and the region of analysis R1. (B) Negative control of the marker. (C-F) FACS analysis of parasites incubated with BSA-Au for periods of 10 min to 2h. The kinetic show that the labelling is time-dependent.*

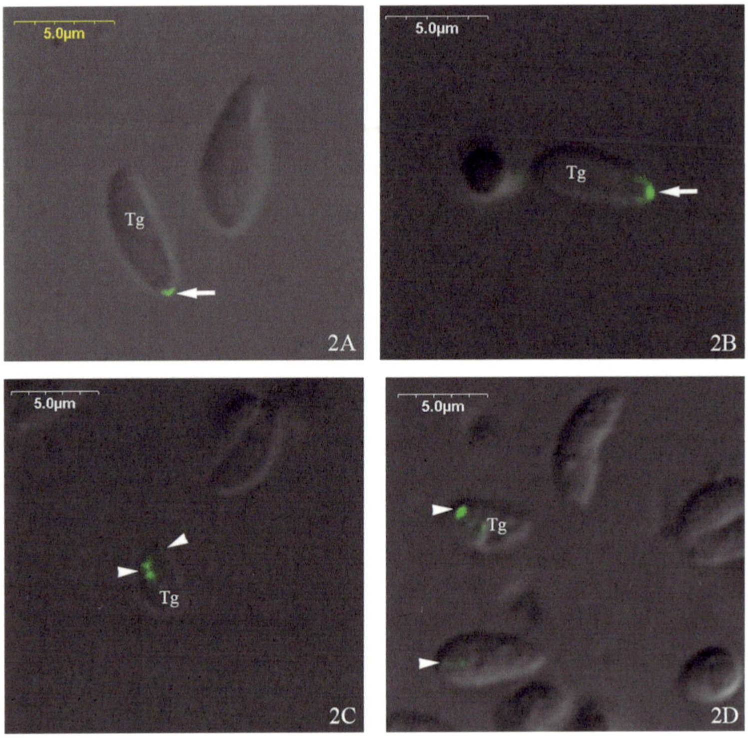

Figure 2.
Confocal microscopy analysis of T. gondii *tachyzoites incubated with Bovine Serum Albumin conjugated with FITC (BSA-FITC). (A and B) At 37 º C for 5 min: (A) the tracer is observed in the apical region (arrow) and (B) a fine granulation extending symmetrically along the first anterior third of the parasite's body. (C) At 37°C for 30 min: the marker is located in the first third of the parasite body (arrowhead). (D) After 2 h at 37°C: BSA-TRICT is internalized through its posterior region (arrowhead).*

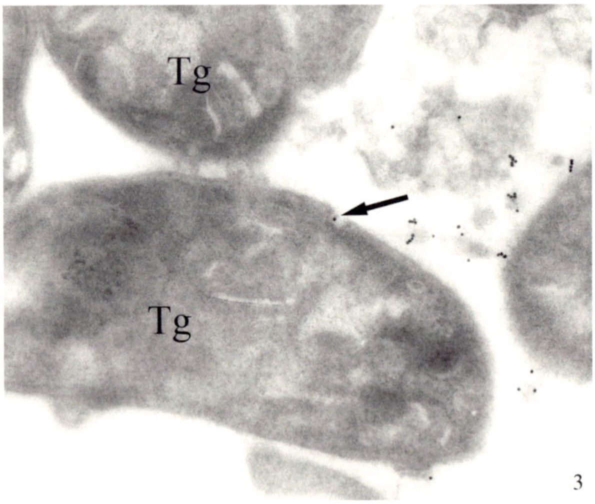

Figure 3.
Ultrastructural analysis of tachyzoite of T. gondii *incubated with BSA-Au. For 10 min. Image showing a gold particle (arrow) at a plasma membrane depression, which displays compatible characteristics such as a micropore.*

3.2 Peroxidase

Experimental assays using HRP as a tracer of fluid phase endocytosis in tachyzoites were performed for transmission electron microscopy analysis only. No labeling was observed in the parasites after incubation at 4°C for 20–30 min (data not shown). The endocytic capacity of tachyzoites was tested during incubation for periods ranging from 5 min to 2 h at 37°C. During the course of the experiments, localization of HRP-Au on the surface of the parasites was rarely observed showing a low association of tracer. The labelling profile was altered when we increased the concentration

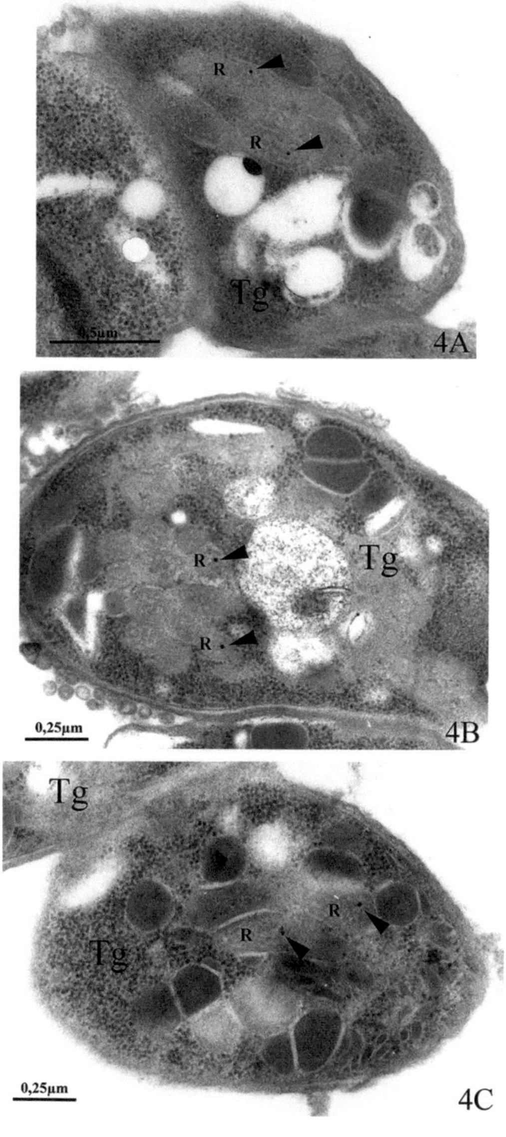

Figure 4.
Ultrastructure of tachyzoites incubated for 5 min to 2 h at 37°C with HRP-Au. Longitudinal sections of parasites revealed particles of the HRP-Au complex (arrowhead) inside rhoptries (R). It was common to observed HRP-Au particles in more than one rhoptries (R) in the same tachyzoite.

of HRP-Au, revealing a greater association of colloidal gold particles on the surface of the tachyzoites (not shown). In all experimental assays performed at 37°C, the intracellular localization of HRP-Au in tachyzoites was exclusively into rhoptries (**Figure 4A–C**). Presence of HRP-Au particles was commonly observed in more than one rhoptry in the same parasite (**Figure 4A–C**).

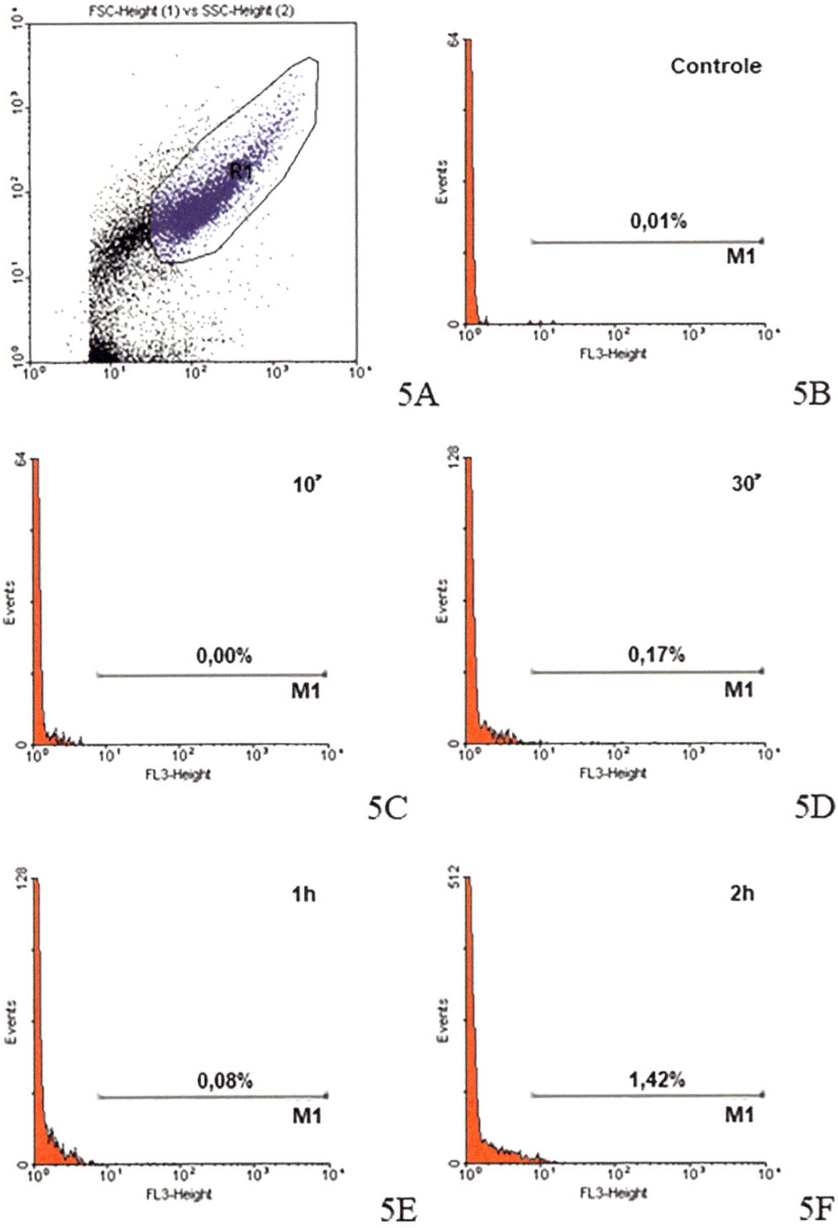

Figure 5.
Analysis by flow cytometry of the incubation of tachyzoites with Dextran-TRITC. A: Tachyzoites incubated with PBS. Graphic show the morphology of the parasites, in terms of size and granularity (FSC x SSC) and the region of analysis R1. (B) Negative control. (C-F) The kinetics of Dextran-TRITC incorporation showed a low percentage of labelled parasites over time

3.3 Dextran

Quantitative analysis of tachyzoites labelling with dextran-TRITC was performed by flow cytometry. The negative control is parasites without prior incubation with the fluorescent tracer (**Figure 5A** and **B**). The results obtained from 10 min to 2 h of incubation at 37°C showed a slight increase in the labelling as a function of incubation time, however with very low values of 0.17% and 1.42% after 30 min and 2 hours, respectively (**Figure 5C–F**). We did not observe any parasite labeled by confocal microscopy.

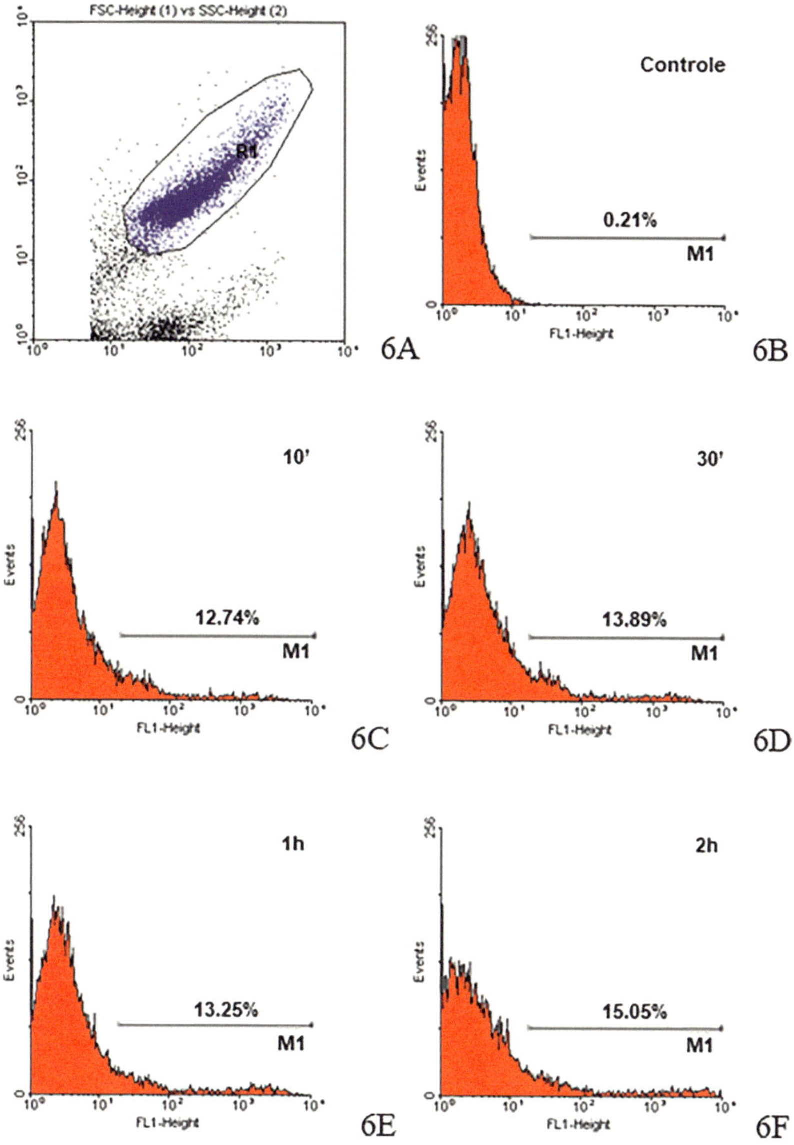

Figure 6.
Representative histograms of flow cytometry showing the kinetic of transferrin conjugated with FITC (Tf-FITC) internalization by T. gondii *tachyzoites incubated at 37°C for different lengths of time. (A) Parasites incubated with PBS. Graphics showing the morphology of the parasites, in terms of size and granularity (FSC x SSC) and the region of analysis R1. (B) Negative control of the marking. (C-F) Kinetics of the incorporation of Tf-FITC by the tachyzoites. The percentage of marked parasites remained constant during incubation for 10, 30, 60 and 120 min.*

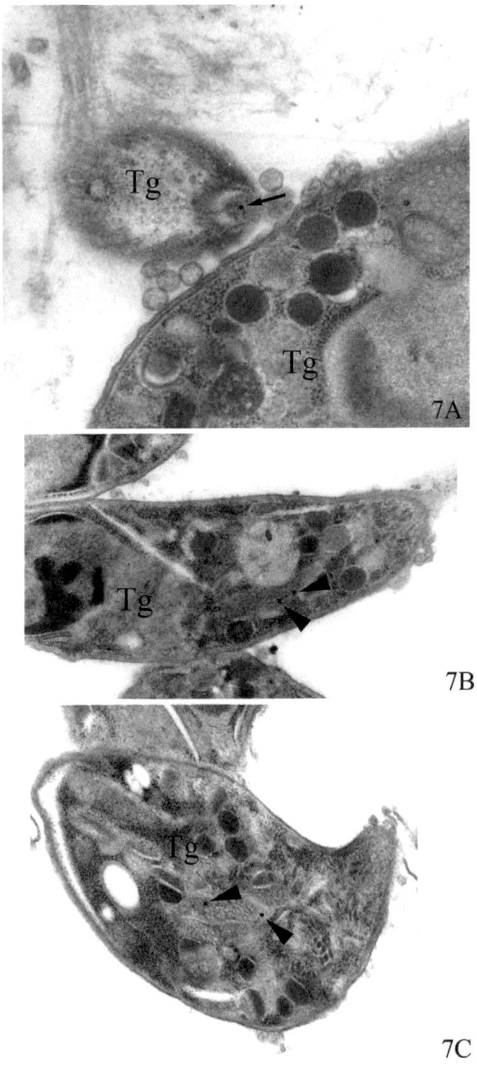

Figure 7.
Ultrastructure of tachyzoites incubated for 5 min to 2 h at 37°C with transferrin-Au (Tf-Au). (A) Transverse/oblique section of the apical region of a tachyzoite with the frontal view of the conoid and the location of the Tf-Au particle at its tip, positioned centrally (arrow). (B and C) Longitudinal ultrafine cut of tachyzoite displays two Tf-Au complexes inside the same rhoptry (arrowhead).

3.4 Receptor-mediated endocytosis

Quantitative analysis of tachyzoites incubated with transferrin was performed by flow cytometry. Negative control is parasites maintained in PBS alone (**Figure 6A** and **B**). After incubation with Tf-FITC, we observed that the number of Tf-FITC-associated tachyzoites was relatively constant reaching levels of 12.74% and 15.05% after 10 min and 2 h, respectively (**Figure 6C** and **F**).

By confocal microscopy analysis after 5 min incubation at 37°C with Tf-FITC we did not observe any tracer-associated parasites. However, 30 min or 2 h incubation at 37° C with Tf-FITC resulted in the uptake of the tracer by the tachyzoites. Tf-FITC was seen concentrated at one pole of the parasite's body or distributed throughout its cytoplasm (data not shown). Ultrastructural analysis of tachyzoites incubated with Tf-Au showed a low association with the surface of the parasites. After incubation for 30 min at 37°C, tachyzoites were observed in transverse/oblique sections of the apical region to allow frontal visualization of the conoid. There was localization of the tracer at the tip of tachyzoites, positioned centrally (**Figure 7B**). These analyzes also revealed the presence of Tf-Au particles inside the rhoptries (**Figure 7B** and **C**).

4. Discussion

The study of nutrient uptake mechanisms by *T. gondii* represents a challenge, mainly because they are mandatorily intracellular parasites. They invade the host cell and replicate delimited within a parasitophorous vacuole (PV), delimited by a host-derived membrane that is extensively modified by the parasite to facilitate nutrient acquisition and minimize attacks from the host cell [14]. This compartmentalization represents physico-biochemical barriers, involving the plasma membrane of the host cell, the membrane of the PV and also the membrane of the parasite. Until recently, little has been explored regarding the pathways of incorporation of macromolecules by tachyzoites. The origin of this knowledge comes mainly from the very early studies on this topic written by: (i) Sénaud et al. (1976) who proposed the micropore as a nutritional organelle [7]; and (ii) Nichols et al. (1994) [8], who used the fluid phase tracer (HRP) to demonstrate the incorporation of cyst matrix material by bradyzoites through the micropore. This process involved the formation of vesicles coated or not by clathrin. Nevertheless, the proposal of the micropore's role as a specialized endocytosis site is still under discussion [6]. Here we investigated the uptake capacity of tachyzoites by using macromolecules routinely employed for endocytosis assays, such as fluid phase and receptor-mediated endocytosis tracers..

The performance of the endocytosis assays, with single parasites through FACS enabled quantitative analysis of labeled parasites in a whole population using different endocytosis tracers. Our results showed that the incorporation or binding was tracer dependent. The highest values by FACS were achieved with BSA as endocytic tracer. Percent of parasites positive for labeling with BSA-FITC increased in time dependent manner from 16.5%, at 10 min to 32%, at 2 hours. In this case, it showed us a time-dependent labeling with the tachyzoites. On the other hand, by using the fluid phase endocytic marker dextran, it turned to be that their association was extremely low, independent of the time of incubation. The concentration of dextran used was not a determining factor of its association with tachyzoites, since using twice the concentration (5 mg/ml) we obtained similar association indexes (1.4% after 2 hours of incubation). They mentioned that only a minority of labeled parasites was found after incubation with transferrin, without having conducted a systematized study. However, results presented here show binding with. This tracer. In this case, the labeling was not time-dependent and reached levels of 15% of labeled parasites.

We tried to infer whether the size of the molecule would affect the ability of tachyzoites uptake. For example, the dextran tracer selected for this study was 4.4 kDA including fluorochrome. The BSA and transferrin tracers have mass approximately 66 kDa and 80 kDa, respectively. Our data demonstrate that the size of the molecule cannot be responsible for the level of incorporation of the marker, considering that the lowest indices obtained in the current work were observed during the assays with dextran, which was the smallest molecule employed by us and that due to its low association with tachyzoites, it was not possible to determine its intracellular location.

It was not possible in our experimental conditions to observe a large number of tachyzoites marked with the tracers used. This difficulty has been noted for fluid and receptor-mediated endocytic markers in the review article by Robibaro et al. (2001) [15]. Results showed that only with high concentrations (around 2 mg/ml) of dextran-FITC and lucifer yellow, allowed detection of uptake by the tachyzoites, allowed to detect uptake by the tachyzoites, reaching percentages of 1 and 8%, respectively, without, however, identifying the route of internalization or the location of the markers in the parasites. Prior data from incubation with BSA-FITC showed a diffuse labeling in some parasites, with no evidence of incorporation via vesicle-like structures.

We also showed the focal BSA-FITC labeling of the surface of the parasite was limited to the anterior surface of the parasite, limited to the anterior (apical) region, where the conoid is located. These data were later corroborated by ultrastructural analyses, where we showed BSA-gold nanoparticles marking the same region of the parasite. The endocytosis tracer was restricted to the two thirds of the anterior of the anterior region of the tachyzoite, contained in elongated vesicles located immediately below the tachyzoite membrane. Additionally, after 2 hours, we documented a diffuse distribution of the marker in the posterior region of some parasites. We demonstrated here that despite the low number of parasites capable of endocytosis with BSA, this kinetic study shows the adhesion of the marker to the surface of the parasite, restricted to the apical region, its subsequent incorporation into structures located in the first two thirds of the parasite body and a possible traffic of this tracer to the posterior region of the parasite.

Aiming to study the receptor-mediated endocytosis pathway in *T. gondii*, we employed transferrin as ligand. Our data suggest that the mechanisms used by Toxoplasma tachyzoites, may be through: (i) the membrane in the posterior third of the parasite body, by the presence of vesicles containing HRP just below the plasma membrane, a region devoid of subpellicular microtubules, which would favor a greater endocytic activity of the membrane; (ii) the membrane in the first anterior third of the parasite body, due to the presence of vesicles and tubules containing HRP particles in this region. Transferrin-specific receptors have been demonstrated in some other protozoans [16–21]. Regarding Plasmodium falciparum, the data are controversial [22–25]. In the case of *T. gondii*, it has been described that lactoferrin, a protein of the transferrin family, binds to its surface [26]. Our quantitative results, through flow cytometry, showed that the association of transferrin with tachyzoites was stable over 2 hours (about 15%). The analysis of this association by confocal microscopy showed the localization of the protein in the interior of the parasites, initially concentrated in the apical region and after 2 hours, located in the median region of its body. *T. gondii* tachyzoites were able to incorporate transferrin suggesting the presence of receptors for transferrin. The presence of transferrin binding sites on the surface of a subgroup of parasites may be related to a certain stage of its cellular cycle, or be dependent on the induction of expression of these surface receptors in the presence of the ligand, as proposed by Botero-Klein et al. (2001) [27], during the characterization of heparin receptors in extracellular tachyzoites. Thus, these data indicate the need for the identification and characterization of this possible receptor for transferrin or an independent iron capture pathway in *T. gondii*. Since there is no excrement route of iron, nor in single - or multicellular organisms, its homeostasis is dependent on the regulation of the level of its uptake, essential for most eukaryotes and prokaryotes [28].

Nichols et al. [8] observed by ultrastructural analysis the internalization, of the fluid phase marker HRP by the micropore of *T. gondii.* Since the base of the micropore in both tachyzoites and bradyzoites is sometimes coated with clathrin, it is possible that receptor-mediated endocytosis via the micropore also occurs. The possible role of the micropore as a site of endocytosis is still unclear and under discussion. Endocytosis of fluid phase tracers via non-specific pinocytosis appears to occur at sites located below the apical region. Our studies by confocal microscopy on the internalization by tachyzoites of fluid phase markers and those of receptor-mediated endocytosis demonstrated the association occurred preferentially through the apical region of the tachyzoites, with their intracellular localization into elongated structures present in the first anterior third of the parasite body. Thus, our results suggest that the micropore is not the only single or preferred route of incorporation of macromolecule by the tachyzoites. The current ultrastructural analysis allowed us to accumulate numerous pieces of evidence of the intracellular presence of endocytic tracers in *T. gondii*.

The question is: How would the markers have access to the interior of the rhoptries? Possibly, by anterior region, via conoid, since it is a region of high exocytic activity allowing the secretion of compartmentalized molecules into the external milieu during the stages of binding (adhesive proteins of the micronemes) and invasion of host cells by parasites (proteins of the rhoptries, involved in the biogenesis of the PVM). It has been proposed that terminal part of the peduncle of the rhoptry fuses with the plasma membrane lining the conoid and the entire surface of the parasite [29] and enables movement of the rhoptry contents from the intracellular to the extracellular environment. It is possible that a reflux of extracellular material may occur into the rhoptries during the exocytosis of thier contents. Based on the evidence presented here, specifically colloidal gold particle presence in the anterior region of the conoid, and accumulation of fluorescent markers at the end of the anterior region of tachyzoites, seen by confocal microscopy (**Figures 1–5**) we hypothesize existence of an inverse pathway where molecules can transit into the rhoptry. Botero-Kleiven et al., (2001) [27], have described that heparin receptors, which we would expect to be located where there is access to host cell were localized in elongated structures, perpendicular to the longitudinal axis of the tachyzoite with a size ranging from 0.5–1.5 nm. This description is compatible with the morphology of the rhoptries, although the authors did not mention this hypothesis.

Another route movement of proteins into rhoptries could be endocytosis via micropore, considering that: (i) previous results have been shown to be a site of endocytosis with vesicles formation at its base either coated or not by clathrin [8]; (ii) there is indirect evidence by confocal microscopy of the presence of heparin binding sites in the anterior lateral region of the parasite body that could correspond to the micropore [27]; (iii) our rare images (**Figure 2A** and **3C**) implicate posterior pore endocytosis by demonstrating the location of markers, after short incubation times, in the posterior region of the parasite, in proximity to the pore (**Figure 6**). This pore has been described as a site of great exocytic activity of dense granules, as demonstrated by Sibley et al. (1995) [27]. Following the same line of reasoning proposed for endocytosis via the conoid, this could also be a site potentially capable of incorporating molecules. The involvement of the posterior pore in this process is supported by the review in Romano and colleagues attributing a multifunction ability to the basal complex (posterior pore) during its participation in host vesicle remodeling and/or lipid uptake [30–32].

Figure 8.
Schematic representation of three different nutrient absorption portals suggested for Toxoplasma tachyzoites: conoid reaching rhoptria, in red; micropore in blue and basal complex in green. The hypothesis is that the uptake of nutrients via the conoid, micropore or basal complex could be a intracellular transported route of nutrients to the rhoptries and be digested there, considering that the pH 5.0 of the rhoptries, is a favorable environment for the activation of lysosomal enzymes

Based on the results presented here and supported in literature we suggest a hypothetical model proposing 3 nutrient uptake portals of macromolecules by tachyzoites of *T. gondii* (**Figure 8**). The first portal (or site) proposes the endocytosis of macromolecules by conoid with its subsequent localization in the rhoptries interior. The second portal, by micropore, leads to vesicle-based transport to rhoptries or to the Golgi following to the rhoptries. Or yet, the third portal would be through the posterior pore with subsequent transport to the Golgi apparatus and/or to the rhoptries. Our hypothesis that the transit of molecules incorporated by *T. gondii* has the rhoptries as its final destination is based on evidence that they are the only acidified organelles of the parasite and that they would be analogous to lysosomal secretory structures which receive material from the endocytic pathways of the cell [33].

In the future, we hope there will be many strategies and approaches to control diseases, like Toxoplasmosis. One Health should bridge disciplines linking human health, animal health, and ecosystem health and that treating and managing Toxoplasmosis demands integrative approaches to breach disciplinary boundaries. Nevertheless, to characterize the mechanisms and portals involved in the acquisition of nutrients by tachyzoites of *T. gondii* and identifying its portals may be an important contribution to the understanding of the biology of the parasite and also be applied as a target for drug action.

5. Conclusions

The ubiquitous parasite *Toxoplasma gondii* has been the subject of intense investigation over the last years. Neverthless, there are still many open questions regarding the nutrient acquisition by this parasite. In this work, we tested the endocytic capacity of extracellular tachyzoites of *T. gondii*. Our results showed innumerous evidence of the intracellular presence of endocytic tracers in the anterior region of the conoid, inside the rhoptries, at the micropore and posterior pore of the parasite. As a result, we proposed a hypothetical scheme (**Figure 8**) aiming to summarize data presented here. These hypotheses should be confirmed through further studies using 3D reconstruction of serial ultrathin sections, for example, or through live cell imaging. The set of data presented here may contribute to new perspectives in this field, enabling in the future a better understanding of the biology of *T. gondii*, enabling application in therapeutic interventions in the treatment of toxoplasmosis.

Acknowledgements

We thank Sandra Maria de Oliveira Souza, Rômulo Custodio dos Santos and Genesio Lopes de Faria for technical assistance. We are grateful also for Carlos Bizarro from Confocal Platform of PDTIS-Fiocruz for help with an Olympus FV 300/BX51 laser scanning confocal microscope and Bruno Avila for the support with the images. This work was supported by grants from Conselho Nacional de Desenvolvimento Científico e Tecnológico (CNPq), Fundação Carlos Chagas Filho de Amparo à Pesquisa do Estado do Rio de Janeiro (FAPERJ), Fundação Oswaldo Cruz (Programa Estratégico de Apoio à Pesquisa em Saúde - PAPES VI), Pronex - Programa de Apoio a Núcleos de Excelência - CNPq/FAPERJ, Edital FAPERJ N° 15/2019 Apoio a Redes de Pesquisa em Saúde no Estado do Rio de Janeiro and Instituto Oswaldo Cruz/Fiocruz.

Conflict of interest

The authors declare no conflict of interest.

Author details

Marialice da F. Ferreira-da-Silva[1], Mauricio Magalhães de Paiva[2], Erick Vaz Guimarães[1] and Helene S. Barbosa[1*]

1 Structural Biology Laboratory, Oswaldo Cruz Institute, Oswaldo Cruz Foundation, Rio de Janeiro, RJ, Brazil

2 Corrosion and Degradation Division National Institute of Technology, Rio de Janeiro, RJ, Brazil

*Address all correspondence to: helene@ioc.fiocruz.br

IntechOpen

© 2022 The Author(s). Licensee IntechOpen. This chapter is distributed under the terms of the Creative Commons Attribution License (http://creativecommons.org/licenses/by/3.0), which permits unrestricted use, distribution, and reproduction in any medium, provided the original work is properly cited.

References

[1] Flegr J, Prandota J, Sovičková M, Israili ZH. Toxoplasmosis--a global threat. Correlation of latent toxoplasmosis with specific disease burden in a set of 88 countries. PLoS One. 2014;**9**(3):e90203

[2] McLeod R, Lykins J, Gwendolyn Noble A, et al. Management of congenital toxoplasmosis. Current Pediatric Report. 2014;**2**:166-194

[3] McAuley JB. Congenital toxoplasmosis. Journal of Pediatric Infectious Diseases Society. 2014;**3**(Suppl 1):S30-S35

[4] Wallon M, Peyron F. Effect of antenatal treatment on the severity of congenital toxoplasmosis. Clinical Infectious Diseases. 2016;**62**(6):811-812

[5] Flegr J. Toxoplasmosis is a risk factor for acquiring SARS-CoV-2 infection and a severe course of COVID-19 in the Czech and Slovak population: A preregistered exploratory internet cross-sectional study. Parasites & Vectors. 2021;**14**(1):508. DOI: 10.1186/s13071-021-05021-9

[6] McGovern KE, Nance JP, David CN, Harrison RES, Noor S, Worth D, et al. SPARC coordinates extracellular matrix remodeling and efficient recruitment to and migration of antigen-specific T cells in the brain following infection. Scientific Reports. 2021;**11**(1):4549. DOI: 10.1038/s41598-021-83952-0

[7] Senaud J. L'ultrastructure du micropyle des Toxoplasmasida [The ultrastructure of the micropyle of Toxoplasmasida]. C R Acad Hebd Seances Acad Sci D. 1966;**262**(1):119

[8] Nichols BA, Chiappino ML, Pavesio CE. Endocytosis at the micropore of *toxoplasma gondii*. Parasitology Research. 1994;**80**(2):91-98. DOI: 10.1007/BF00933773

[9] Roth TF, Porter KR. Yolk protein uptake I the oocyte of the mosquito *Aedes aegypti*. L. Journal of Cell Biology. 1964;**20**(2):313-332. DOI: 10.1083/jcb.20.2.313

[10] Dlugonska H. Toxoplasma rhoptries: Unique secretory organelles and source of promising vaccine proteins for immunoprevention of toxoplasmosis. Journal of Biomedicine & Biotechnology. 2008;**2018**:632424

[11] Saliba KJ, Kirk K. Nutrient acquisition by intracellular apicomplexan parasites: Staying in for dinner. International Journal for Parasitology. 2001;**31**(12):1321-1330

[12] Carvalho L, de; Souza, W. De. Cytochemical localization of plasma membrane enzyme markers during interiorization of tachyzoites of *toxoplasma gondii* by macrophages. The Journal of Protozoology. 1989;**36**:164-170

[13] Frens G. Controlled nucleation for the regulation of the particle size in monodisperse gold suspensions. Nature Physical Science. 1973;**241**:20-22

[14] Robibaro B, Stedman TT, Coppens I, Ngô HM, Pypaert M, Bivona T, et al. *Toxoplasma gondii* Rab5 enhances cholesterol acquisition from host cells. Cellular Microbiology. 2002;**4**(3):139-152

[15] Coppens I, Opperdoes FR, Courtoy PJ, Baudhuin P. Receptor-mediated endocytosis in the bloodstream form of *Trypanosoma brucei*. The Journal of Protozoology. 1987;**34**(4):465-473

[16] Schell D, Evers R, Preis D, Ziegelbauer K, Kiefer H, Lottspeich F, et al. A transferrin-binding protein of *Trypanosoma brucei* is encoded by one of the genes in the variant surface glycoprotein gene expression site. The EMBO Journal. 1991;**10**(5):1061-1066

[17] Soares MJ, de Souza W. Endocytosis of gold-labeled proteins and LDL by *Trypanosoma cruzi*. Parasitology Research. 1991;**77**(6):461-468

[18] Voyiatzaki CS, Soteriadou KP. Identification and isolation of the *leishmania* transferrin receptor. The Journal of Biological Chemistry. 1992;**267**(13):9112-9117

[19] de Figueiredo RC, Soares MJ. Low temperature blocks fluid-phase pinocytosis and receptor-mediated endocytosis in *Trypanosoma cruzi* epimastigotes. Parasitology Research. 2000;**86**(5):413-418. DOI: 10.1007/s004360050686

[20] Porto-Carreiro I, Attias M, Miranda K, De Souza W, Cunha-e-Silva N. *Trypanosoma cruzi* epimastigote endocytic pathway: Cargo enters the cytostome and passes through an early endosomal network before storage in reservosomes. European Journal of Cell Biology. 2000;**79**:858-869. DOI: 10.1078/0171-9335-00112

[21] Haldar K, Henderson CL, Cross GA. Identification of the parasite transferrin receptor of *plasmodium falciparum*-infected erythrocytes and its acylation via 1,2-diacyl-sn-glycerol. Proceedings of the National Academy of Sciences of the United States of America. 1986;**83**(22):8565-8569

[22] Pollack S, Schnelle V. Inability to detect transferrin receptors on *P. falciparum* parasitized red cells. British Journal of Haematology. 1988;**68**(1):125-129

[23] Affonso AL, Benchimol M, Ribeiro KC, Lins U, De Souza W. Further studies on the endocytic activity of *Tritrichomonas foetus*. Parasitology Research. 1994;**80**(5):403-413

[24] Tachezy J, Kulda J, Bahníková I, Suchan P, Rázga J, Schrével J. *Tritrichomonas foetus*: Iron acquisition from lactoferrin and transferrin. Experimental Parasitology. 1996;**83**(2):216-228

[25] Tanaka T, Abe Y, Kim WS, Xuan X, Nagasawa H, Igarashi I, et al. The detection of bovine lactoferrin binding protein on *toxoplasma gondii*. The Journal of Veterinary Medical Science. 2003;**65**:1377-1380

[26] Botero-Kleiven S, Fernández V, Lindh J, Richter-Dahlfors A, von Euler A, Wahlgren M. Receptor-mediated endocytosis in an apicomplexan parasite (*toxoplasma gondii*). Experimental Parasitology. 2001;**98**(3):134-144

[27] Kaplan J. Mechanisms of cellular iron acquisition: Another iron in the fire. Cell. 2002;**111**(5):603-606

[28] Sibley LD. Intracellular parasite invasion strategies. Science. 2004;**304**(5668):248-253

[29] Sibley LD, Niesman IR, Parmley SF, Cesbron-Delauw MF. Regulated secretion of multi-lamellar vesicles leads to formation of a tubulo-vesicular network in host-cell vacuoles occupied by *toxoplasma gondii*. Journal of Cell Science. 1995;**108**(Pt 4):1669-1677

[30] Romano JD, Nolan SJ, Porter C, Ehrenman K, Hartman EJ, Hsia RC, et al. The parasite toxoplasma sequesters diverse Rab host vesicles within an intravacuolar network. The Journal of Cell Biology. 2017;**216**:4235-4254. DOI: 10.1083/jcb.201701108

[31] Gubbels MJ, Ferguson DJP, Saha S, Romano JD, Chavan S, Primo VA, et al. Toxoplasma gondii's basal complex: The other apicomplexan business end is multifunctional. Frontiers in Cellular and Infection Microbiology. 2022;**12**:882166. DOI: 10.3389/fcimb.2022.882166

[32] Ngô HM, Yang M, Joiner KA. Are rhoptries in apicomplexan parasites secretory granules or secretory lysosomal granules? Molecular Microbiology. 2004;**52**(6):1531-1541. DOI: 10.1111/j.1365-2958.2004.04056.x

[33] Aguirre AA, Longcore T, Barbieri M, Dabritz H, Hill D, Klein PN, et al. The one health approach to toxoplasmosis: Epidemiology, control, and prevention strategies. Ecohealth. 2019;**16**(2):378-390

Chapter 5

Neuroimmunopathology in Toxoplasmic Encephalitis

Gungor Cagdas Dincel, Abdulaziz Alouffi, Ebtesam M. Al Olayan, Guillermo Tellez-Isaias and Saeed El-Ashram

Abstract

Toxoplasma gondii is a zoonotic protozoan parasite that causes mortality because of significant neuropathology. It is widespread in neonatal infections. Although the neuroimmunopathogenesis of toxoplasmic encephalitis (TE) has been studied for many years, it is still not completely understood, showing the disease's severity. The urge to write this chapter comes at this stage. The sections covered in this chapter show the pathogenesis that has been established and characterized so far. The involvement of astrocytes and microglia in the development of neuropathology, which begins with tachyzoites crossing the blood-brain barrier during acute infection, has been explored. The molecular mechanism between schizophrenia and TE has been thoroughly proven. Uncovering the molecular pathogenesis of TE is critical for both understanding neuropathology and elucidating the link between neuropsychiatric diseases. Each part covered here is expected to contribute to developing novel therapeutic agents for the treatment and maybe prevention of neuropathology. The pathogenesis of the steady progression of encephalitis has been meticulously revealed. Thus, this chapter will offer significant insight into developing novel treatments for all organisms suffering from this disease.

Keywords: *T. gondii*, immunopathogenesis, neuropathology, toxoplasmic encephalitis, cerebral toxoplasmosis

1. Introduction

Toxoplasma gondii (*T. gondii*), an obligate intracellular protozoan parasite, is closely related to public health because its zoonotic nature infects all warm-blooded animals [1, 2]. The severity of *T. gondii* infections is directly proportional to the infected host's immunity level. While the infection is subclinical in immunocompetent people, it causes lethal toxoplasmic encephalitis (TE) in immunocompromised people because of tissue cyst reactivation [3, 4]. *T. gondii*-related neuropathology is not only restricted to TE. *T. gondii* infections have been linked to neuropsychiatric and behavioral problems, including schizophrenia and bipolar disorders, as well as significant mental illnesses, including depression and obsessive-compulsive disorder [5–7]. Even though

 IntechOpen

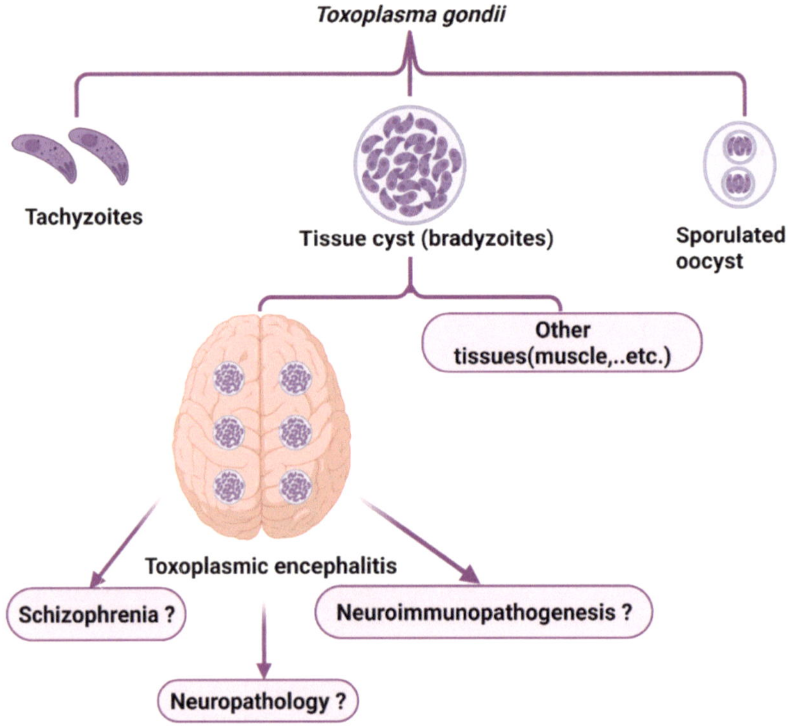

Figure 1.
The association between schizophrenia and schizophrenia mediated by T. gondii.

neuroimmunopathogenesis and neuropsychiatric disorders in TE and the pathogenesis of neuropsychiatric disorders have been studied for a long time, they are still among the issues that are still unclear and have many questions (**Figure 1**).

This section will detail the immunopathogenesis of TE and examine potential mechanisms. Thus, it will shed light on the researchers' goals about what should be considered in the management of the TE process and what should be focused on in future studies. In addition, the intricate criteria needed for the differential diagnosis will be thoroughly covered. Therefore, considering the neuroimmunopathogenesis of the disease and the diagnosis concurrently will facilitate a better understanding of TE.

Tachyzoites, the active life form of *T. gondii*, cannot be entirely eliminated in the brain, unlike in other organs, despite activating a robust immune system upon reaching the brain. At this stage, they convert into a bradyzoite form inside a tissue cyst and continue to grow as a chronic infection, presenting no symptoms throughout the host's life [2, 8]. To reach the brain, this tissue cyst stage must first cross the blood-brain barrier (BBB). This point, the BBB's transitional phase, will be considered the first step (**Figure 2**).

2. Why is toxoplasmic encephalitis so important?

Toxoplasmosis is far more dangerous in immunocompromised people than in healthy people. Multiple organ involvement is possible in cases of acute toxoplasmosis

DOI: http://dx.doi.org/10.5772/intechopen.109341

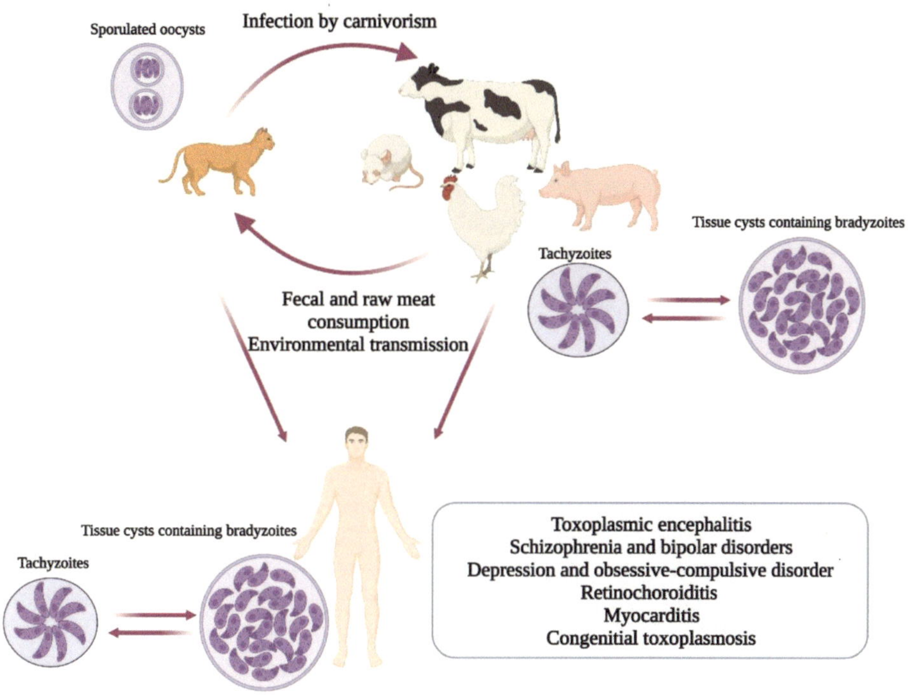

Figure 2.
The T. gondii life cycle. There are two phases in the life cycle: sexual in the definitive host and asexual in intermediate hosts. Unsporulated oocysts are shed in the cat's feces. Intermediate hosts get infected after eating contaminated food or drink. Oocysts evolve into tachyzoites in the small intestine after consumption. When the host's immune system is compromised, tachyzoites induce the acute stage of infection and develop into tissue cyst bradyzoites. Bradyzoites may either remain dormant for the host's life or convert to tachyzoites. Consuming undercooked meat with tissue cysts is one of the primary transmission routes in humans. Clinical signs of the disease include encephalitis, schizophrenia, bipolar disorders, depression, obsessive-compulsive disorders, retinochoroiditis, myocarditis, and fetal abnormalities following transplacental infection in immunocompromised people.

in immunocompromised people. Systemic diseases are often observed with severe pneumonia and retinochoroiditis, but encephalitis is the most common clinical finding [9]. The neuroimmunopathogenesis should be fully explained at this point. There is evidence that toxoplasmosis can be reactivated in people with certain cancer types, including lymphoma, leukemia, and myeloma [10–12]. Organ transplantation also poses a significant risk of toxoplasmosis, which can be fatal. For example, transplanting a *T. gondii*-infected organ to an immunocompromised patient may reactivate the organ's latent infection. Furthermore, reactivating latent *T. gondii* infection because of immunosuppressive treatment after organ transplantation can cause a fatal condition [13]. Importantly, deaths have been reported in Acquired Immune Deficiency Syndrome (AIDS) patients with CD4 T lymphocyte cell counts <200 cells/μL and severe immunosuppression, resulting in the formation of reactivation-related TE [14, 15]. It has been reported, for example, that TE develops in approximately 30–40% of immunocompromised *T. gondii* seropositive ADIS individuals [16]. It is clear that this rate is extremely high. Actually, TE is quite crucial for patients of all ages. To differentiate between patients, it is evident that immunocompromised individuals are at a higher risk. A thorough analysis of the disease's molecular pathogenesis is required at this point. Otherwise, deaths caused by TE make us to disregard the primary disease.

3. Crossing of the blood-brain barrier by *T. gondii* and molecular mechanism of damages

Although the blood-brain barrier (BBB) is a solid barrier between blood and brain tissue, *T. gondii*'s superior tactics for crossing this barrier and reaching the brain have yet to be completely disclosed. However, it is still unclear whether neuroinvasion occurs via hiding within or as a direct tachyzoite outside the cell. The actual huge issue begins right here. Because TE, one of the most severe forms of the disease that may lead to death, develops when tachyzoites cross through the properly functioning BBB during acute infection [14]. The BBB is lined by microvascular endothelial cells, which are highly specialized. The key characteristics of this barrier are low pinocytic activity, restricted fenestration, and high trans-endothelial resistance. Thus, it functions as a structural and functional barrier with minimal permeability. Additionally, pericytes, microglia, and astrocytes are closely associated with BBB endothelial cells and substantially contribute to sustaining BBB activity [17–20]. It has been proven conclusively that the first transmission of *T. gondii* to the central nervous system (CNS) occurs primarily through the cortical capillaries [21]. It is crucial at this point to explore how this dissemination occurred. There is evidence that *T. gondii* manipulates gene expression in brain endothelial cells to cross the BBB via a "Trojan horse" strategy. *T. gondii* in cells expressing CD11b may play a significant role in the intracellular trafficking of the BBB [22]. In the same research, the percentage of CD45+/CD11bc+ cells infected with *T. gondii* tachyzoites was 13 times greater at the beginning of infection when compared with the population percentages. In this work, CD45+/CD11bc+ cells, both infected and uninfected, show the ability to breach the BBB in an in vitro culture model. Uninfected cells respond identically to infected cells, which is a surprise topic of debate at this level. Similar increases in the expression of intercellular adhesion molecule 1 and monocyte chemotactic protein-1 were reported after infection of brain endothelial cells with *T. gondii* tachyzoites within 2 and 12 hours, respectively [22]. CD11c- CD11b+ monocytes and CD11c+ CD11b+/– dendritic cells are the progenitor cells that facilitate the switch to the "Trojan horse" approach [23]. Although extracellular free-moving tachyzoites can cross the BBB without an intermediate, they can also do so *via* a variety of other mechanisms. Infection of the microvascular endothelium of the brain by tachyzoites, and their subsequent reproduction in the CNS is one potential approach [24]. Toxophyllin and protease expressions have been observed to assist the passage of *T. gondii* through the BBB. It has been shown that this transition is accomplished by destroying BBB cells [23, 25]. Because of this destruction, the BBB's permeability and integrity have been disrupted. Importantly, BBB disruption induces irreversible neurodegeneration and neuropathology in TE. In TE, *T. gondii* directly damages the blood-brain barrier (BBB) by inducing apoptosis in endothelial cells [26], while *T. gondii*-mediated oxidative stress (OS) [27] and nitrosative stress (NS) [28] also induce the same damage. Therefore, it is crucial to uncover the intricate processes involved in the pathophysiology of BBB degradation. It has been shown that von Willebrand Factor, closely related to a disintegrin and metalloprotease with thrombo-spondin 1 repeats family 13 (ADAMTS-13), plays an essential role in BBB permeability [29]. In addition, it has been underlined that ADAMTS-13 has neuroprotective roles by disclosing and decreasing BBB damage in the brains of small ruminants attacked with border disease virus [30] and listeric encephalitis [31]. Similarly, the expression of ADAMTS-13 has strongly shown BBB destruction in TE, and its neuroprotective activities have been described [26]. These results show that ADAMTS-13 has essential functions in the

BBB and neuronal parenchyma in disclosing the degree of BBB damage and reducing neuropathology. This essential initial step must be fully disclosed. Because a thorough understanding of the processes by which tachyzoites cross the BBB will aid in developing more focused therapies for preventing fatal brain injury. The primary aim should be to maintain the integrity of the BBB and prevent tachyzoites from entering the brain parenchyma during the acute infection phase. Thus, chronic TE may be avoided even before it begins.

4. Molecular mechanism of neuropathology caused by *T. gondii*

Tachyzoites that reach the brain parenchyma after improved strategies have crossed the BBB may induce severe neuropathology and host mortality. Although the molecular pathways of neuropathology are continuously investigated, there are still several unanswered questions. Chemokine (C–C motif) receptor 2 (CCR2), a crucial chemokine receptor with a microbicidal function, has a tight relationship with the parasite burden in the brain. In mice whose CCR2 receptor was experimentally blocked, the immune system cells in the brain were markedly inactive, resulting in a considerable increase in the parasite load in the brain [32]. In conclusion, the CCR2 receptor is necessary to regulate parasite replication in peripheral organs and the CNS in particular (**Figure 3**). Although the deadly TE produced by the reactivation of latent tissue cysts in the brain has been investigated for many years, the molecular mechanism of reactivation is still unknown.

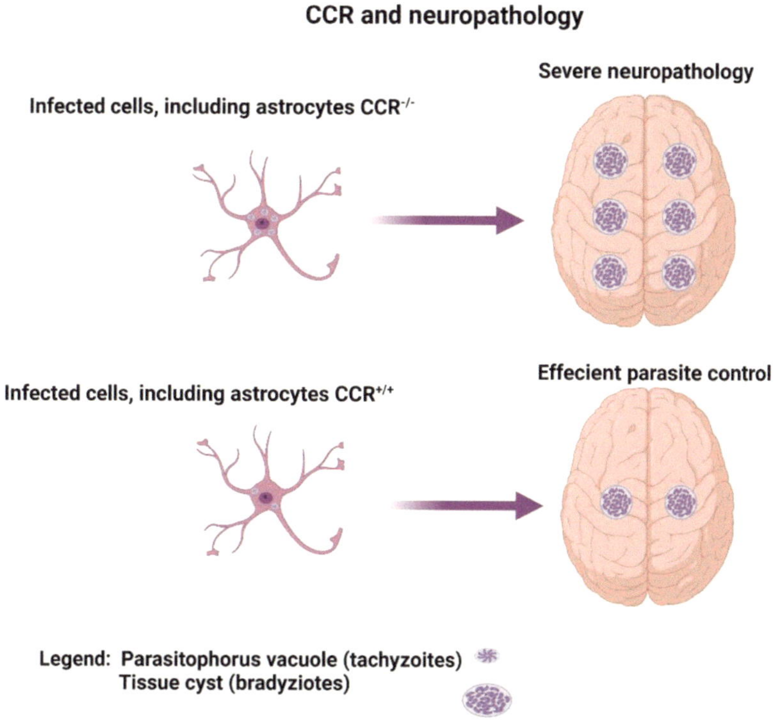

Figure 3.
The role of CCR2 in TE.

The significance of nitric oxide (NO) generation in chronic TE is remarkable since it must be in a delicate balance. This section focuses on two distinct topics. The first is NO's neuroprotective impact, while the second is its neurotoxic and neuropathological consequences, which may be severe. During *T. gondii* infection, nuclear factor kappa B and inflammatory cytokines are overexpressed in blood-derived macrophages. IL-1α was overexpressed in microglia and IL-1β in macrophages during infection compared with control groups. In TE, IL-1R1, gasdermin-D-dependent IL-1α, and caspase-1/11 have been demonstrated to control parasite multiplication, limit neuroinflammation, and regulate immune cell multiplication infiltration [33]. This research shows that alarmin IL-1α, produced by microglia, works hard to reduce the neuropathology found in TE (**Figure 4**).

The anti-parasitic action of NO is diminished in C57BL/6 mice vulnerable to chronically infected *T. gondii*, and TE, which leads to mortality, occurs because of the reactivation of tissue cysts [34–38]. NO plays a significant role in the shift from acute to chronic infection. There have been investigations into the critical functions of NO in the shift from acute to chronic infection. In a nutshell, NO causes the conversion

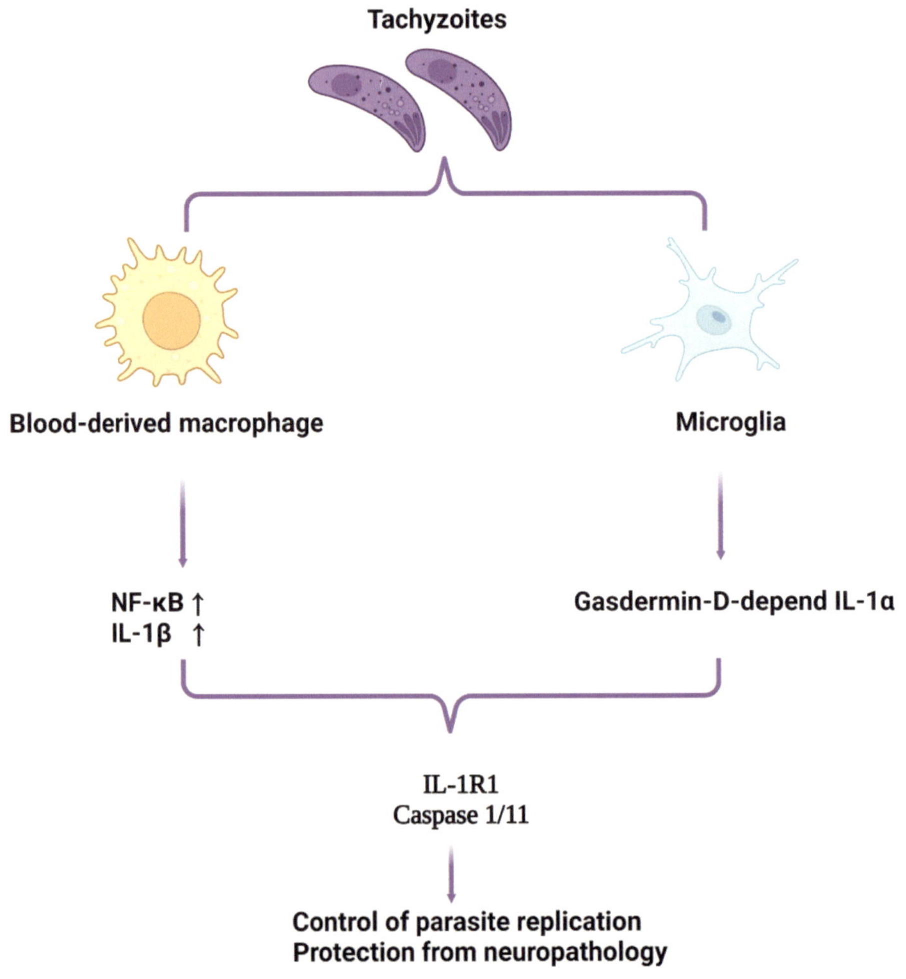

Figure 4.
IL-1R1 and Caspase 1/11-mediated pathway for TE attenuation.

of tachyzoite to bradyzoite forms and attempts to regulate the course of the illness [34–36, 38–43]. It has been shown that a reduction in inducible nitric oxide synthase (iNOS) expression is linked to tissue cyst reactivation [34, 37, 38]. However, the focus of this section will be on the occurrence of neuropathology induced by NO generated over physiological limitations, because it has been proven that NO causes significant neuropathology rather than having neuroprotective characteristics. Despite the emphasis on the consequences of iNOS-derived NO generation in TE, it was shown that NO was not only iNOS-derived. Endothelial nitric oxide synthase (eNOS)-derived NO produced from endothelial cells infected with tachyzoites, and neuronal nitric oxide synthase (nNOS)-derived NO expressed by neurons also induces neuropathology at an unavoidable level [28]. It has been shown that all NO sources play a part in disease pathogenesis, explaining why the NS has progressed to such a severe level (**Figure 5**).

Apoptosis causes neuropathology found in TE, as shown by high caspase 3 expression and suppression of B-cell lymphoma-extra-large (Bcl-xL) expression, which has anti-apoptotic characteristics. The observed apoptosis is driven both extrinsically by the expression of Tumor necrosis factor receptor 1 and caspase 8 and internally by the production of caspase 9, which is a component of the apoptosome complex. *T. gondii* inhibiting Bcl-xL is exceptionally notable. The most surprising discovery in this research is that Purkinje cells were susceptible to intrinsically caused apoptosis. Thus, intrinsically caused apoptosis has been identified as one reason for Purkinje cell death in TE patients [26]. Pathological neurofilament (NF) accumulation in neurodegenerative cerebral diseases, particularly in injured neurons, is the most fundamental sign of acute neuroparenchymal destruction [44–46]. The critical element to understand in this section is that NF accumulates to pathological levels during the acute stage. TE had a large amount of

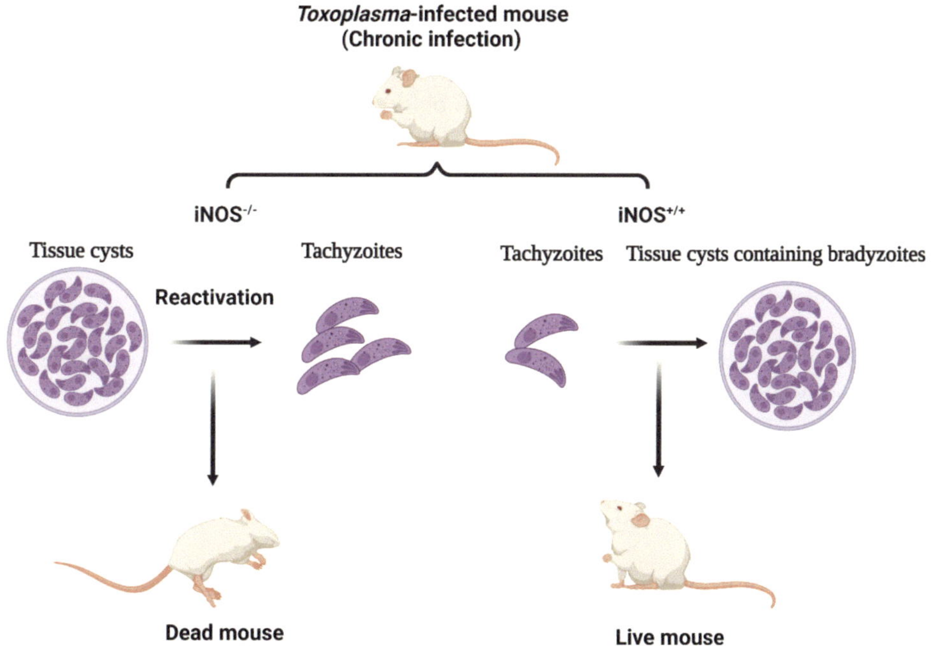

Figure 5.
The role of iNOS in TE survival.

NF buildup in the chronic TE stage. The occurrence of degeneration and necrosis in acute neuroparenchymal structures owing to tissue cyst reactivation was the cause. As a result, it has been shown that monitoring NF accumulations may evaluate and track neuropathology in acute TE. Consequently, it has been demonstrated that the degree of neuropathology in acute TE may be identified and tracked by monitoring NF accumulations [28]. In experimental models of ischemia in rats, neuron-specific enolase (NSE) may be employed as a quantifiable marker to determine the degree of neuronal injury [47]. NSE is a good-quality marker for assessing TE-related neuropathology's severity and disease follow-up. NSE strongly showed the severity of *T. gondii*-mediated neurodegenerations in neuroparenchymal structures [27]. This is a significant result because, given the absence of a successful therapy for TE that has yet to be addressed entirely, it will play an essential role in assessing neuroprotective drugs.

It is widely established that oxidative stress (OS) plays a significant part in the molecular process of neurodegeneration/neuropathology in TE. Glutathione reductase and 8-hydroxy-2′-deoxyguanosine have been expressed at pathological levels in TE, and the expression of Cu/Zn superoxide dismutase (SOD), a critical endogenous enzymatic antioxidant that protects cells against OS-related apoptosis, is similarly hindered. Therefore, OS and n/mt DNA damage produced by OS shows severe neuropathology mediated by *T. gondii*, because a reduction in antioxidant enzyme activity provides crucial insight into the neuropathogenesis of TE [27]. Glia maturation factor (GMF)-mediated proinflammatory responses have been found to play a crucial role in the neuropathogenesis of neurodegenerative and demyelinating diseases and produce severe neuropathology [48–51]. In addition, it has been shown that elevated GMF expressions play a crucial role in the neuropathogenesis of Alzheimer's disease, a severe neurodegenerative disorder [48, 52]. Dincel's 2017 research found abnormal levels of GMF-b expression, a severe proinflammatory cytokine, in reactive glial cells and specifically in gliosis foci in the brain tissues of animals. This research showed a relationship between GMF overexpression in glial cells in TE and neuronal damage. It has been explained that the subsequent neurotoxicity is GMF-mediated neuropathology that has not been documented in TE before [53]. Uncovering a novel GMF-mediated proinflammatory mechanism is a significant step forward to our knowledge of neuropathogenesis.

The pathogenesis of TE has been linked to an enhanced unfolded protein response and extended ER stress. In the model created with *T. gondii*, type I strains showed that Rhoptry protein 18 (ROP18) is associated with the N-terminal portion of endoplasmic reticulum (ER)-associated protein called reticulon 1-C (RTN1-C), an ER protein expressed in the CNS [54]. As a result of ROP18 phosphorylation of RTN1-C, it has been explained that it triggers ER-stress-mediated apoptosis in neuronal cells. In addition to these findings, it has been demonstrated that ROP18 phosphorylation of RTN1-C increases glucose-regulated protein 78 acetylation by decreasing the activity of histone deacetylase. These results have been associated with neuronal apoptosis [54]. These findings clearly show that ER stress and ROP18 expression play critical roles in the pathogenesis of TE, and treatment modalities should focus on this target (**Figure 6**).

The neuropathology found in TE is critical. Although the neuropathology associated with chronic toxoplasmosis is thought to be resolved, we should not ignore their neuropsychiatric consequences. Any reactivation of tissue cysts intimately linked to the immune system might end in mortality. Although the aim here is to reduce neuropathology and avoid permanent damage, we should not ignore that the goal is

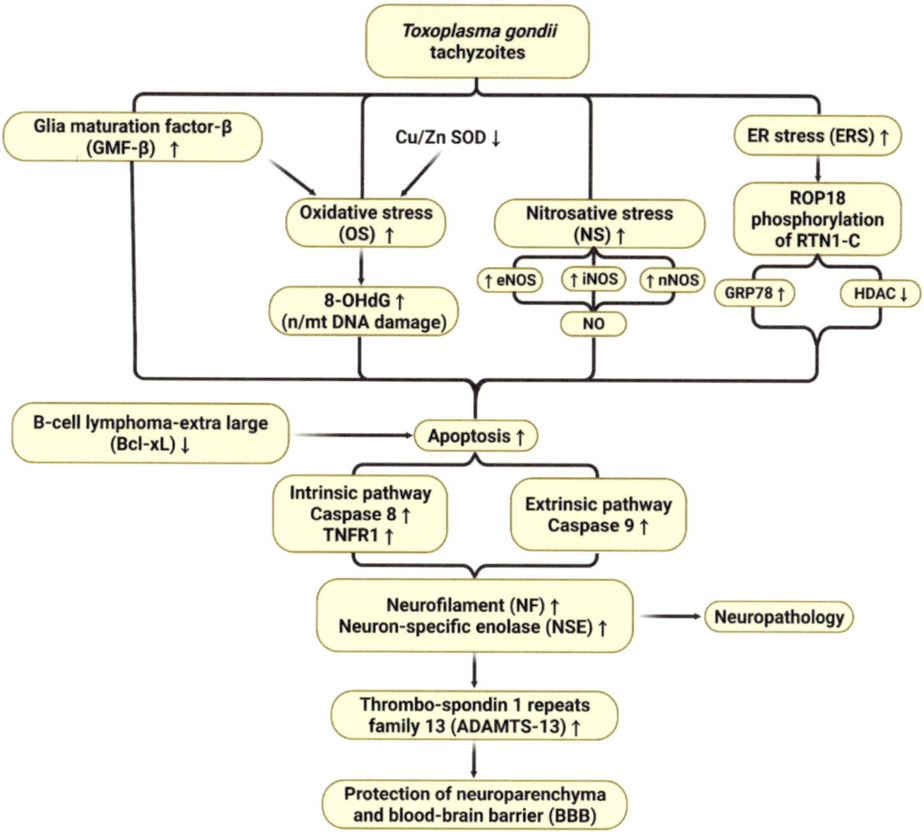

Figure 6.
Mechanism underlying the significant apoptosis seen in TE.

to eliminate tissue cysts. Because as long as the existence of tissue cysts in the brain persists, patients cannot avoid the risk of death.

5. Critical roles of astrocytes in toxoplasmic encephalitis

Astrocytes exert considerable effort to maintain control over *T. gondii*. Although the expression of gamma interferon (IFN-y) has been proven to significantly regulate *T. gondii* in the brain, the molecular mechanism behind these protective actions is not well known. A comprehensive analysis revealed that IFN-gamma-induced GTPase (IGTP) suppressed the replication of *T. gondii* tachyzoites in astrocytes. Therefore, IGTP protects TE *via* astrocyte-mediated mechanisms [55]. However, it has been established that the signal transducer and activator of transcription 1 are required for IFN-gamma to suppress parasite multiplication in astrocytes (STAT1). Experimentally, it has been shown that in animals devoid of STAT1 expression, astrocytes cannot suppress *T. gondii* replication and, thus, cannot prevent neuropathology [56]. It has been shown that the orphan nuclear receptor regulates STAT1 signaling and host defense in astrocytes. In TE, astrocytes play a crucial role in developing a robust resistance to the parasite and complicated molecular pathogenesis [57]. It has been discovered that astrocytes' glycoprotein 130 (gp130) expression plays a crucial role in mice with the experimental TE model. An experiment with animals lacking gp130 showed that apoptosis in astrocytes increased, and inflammatory lesions could not be avoided. However, it has been shown that gp130 expression by astrocytes is not required for *T. gondii* regulation in these cells. In summary, it has been shown that astrocyte-expressed gp130 alleviates neuropathology but does not prevent parasite replication [58].

The importance of astrocytes in the immunopathogenesis of TE cannot be overstated. Astrocytes are among the cells infected by tachyzoites in both *in vivo* and *in vitro* experiments. *In vivo* investigations have shown that tachyzoites target and infect astrocytes after invading the brain. Activated astrocytes have been demonstrated to decrease parasite multiplication in TE and, as a result, help alleviate severe neuropathology such as necrosis [59–63]. Mice have been demonstrated to have little chance of surviving in TE, which was established in mice whose glial fibrillary acidic protein (GFAP) expressions were suppressed, and astrocyte numbers were lowered in an experiment [58]. This condition may be described as follows: in mice without a protective astrocyte population, the efficiency of GFAP, an essential immunoregulatory, would be reduced. As a result, parasite replication rises, and because of this replication, more widespread and severe inflammatory lesions develop, resulting in increased neuropathology severity [58, 64]. Severe GFAP expressions from astrocytes were identified in a potential tissue cyst reactivation. This robust expression shows that astrocytes exert effort to prevent neuropathology in the neuroparenchyma. Therefore, it has been shown that astrocytes function as immunomodulatory and immune effector cells in TE [28].

It has been revealed that astrocytic transforming growth factor-beta (TGF-β) signaling is crucial for suppressing the neuroinflammatory response in TE. It has been shown that inhibiting astrocytic TGF- β signaling increases immune cell infiltration into the brain, resulting in severe neuronal damage. Inhibiting astrocytic TGF-β signaling has no effect on CNS parasite burden in acute or chronic phases, suggesting a separate molecular mechanism in astrocytic TGF-β-mediated neuroinflammation. Astrocytic TGF-β signaling has a role in avoiding neuronal tissue damage in TE *via* astrocytic TGF-β signaling [65].

T. gondii-infected astrocytes have been shown to express prostaglandin E2 (PGE2). PGE2 production by infected astrocytes has also been linked to microglia IL-10 expression reliant on cyclic adenosine monophosphate (cAMP). Finally, *T. gondii*-infected astrocytes suppress NO generation by IFN-gamma-activated microglia, and cAMP-dependent IL-10 expression by microglia contributes to neuropathology reduction (**Figure** 7) [66].

6. *T. gondii* and schizophrenia: Pathogenesis relationship

The relationship between *T. gondii* and schizophrenia has been investigated for decades. However, despite such studies and extensive meta-analyses, the relationship between the two diseases remains obscure. Because it is a current topic, this is regarded as a significant point. Kynurenic acid (KYNA), a metabolite of the kynurenine pathway derived from the breakdown of tryptophan, is synthesized by astrocytes [67–69]. In the pathogenesis of schizophrenia, impaired metabolism of the brain kynurenine pathway (KP) and consequently elevated kynurenate levels have been conclusively demonstrated [70]. In response to intense glial activation during TE, KYNA formation increases substantially. To summarize,

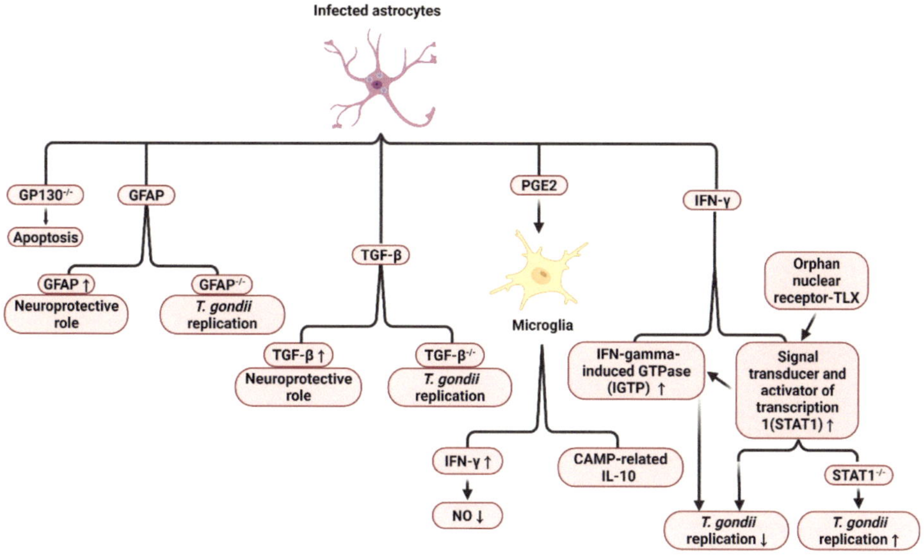

Figure 7.
TE severity and the roles of infected astrocytes.

KP metabolites play an essential role in the pathogenesis of *T. gondii*-mediated schizophrenia (**Figure 8**) [71].

The Matrix Metalloproteinase-9 (MMP-9) gene has been extensively studied in schizophrenia, and it has been demonstrated that the MMP-9 gene polymorphism may play a role in the pathogenesis of schizophrenia [72–74]. In TE, tachyzoite-infected astroglial cells increase the expression of Matrix Metallopeptidase-2 and -9 (MMP-2 and MMP-9). It has even been reported that the inflammatory development of encephalitis can be controlled by inhibiting MMP-2 and MMP-9 expression [75], because MMP-2 and MMP-9 expressed by astroglial cells have been shown to contribute to extracellular matrix degradation in brain tissue [76]. These findings show that MMP expression is essential in the pathogenesis of *T. gondii*-mediated schizophrenia (**Figure 9**).

The similarities between the pathogenesis of schizophrenia and TE are quite striking. There is an increase in NO production in schizophrenia [77] and neuropathology associated with OS [78, 79]. However, it has been demonstrated that the antioxidant enzyme SOD in the brain is significantly decreased in schizophrenia [80]. A considerable drop in the expression of the anti-apoptotic protein Bcl-2 has been discovered in the pathogenesis of schizophrenia [81], and apoptosis has been proven to play a crucial part in the process [82–84]. Similarly, cytokine-mediated neuronal damage has been documented in schizophrenia, and declines in the number and density of neurons have been reported [85]. When examining the pathogenesis of TE considering these data, it has been discovered that there are remarkable similarities, and their link is highlighted. Pathological levels of eNOS, iNOS, and nNOS-derived NO production have been observed in TE [28]. Moreover, it was shown that OS increased while SOD activity reduced considerably [27]. In addition, severe inhibition of the expression of Bcl-xL, an antiapoptotic protein, and triggering of apoptosis have been clearly demonstrated in TE models [26]. In TE, elevated GMF expressions, which induce proinflammatory responses in glial cells, revealed cytokine-mediated neuropathology [53] (**Figure 10**).

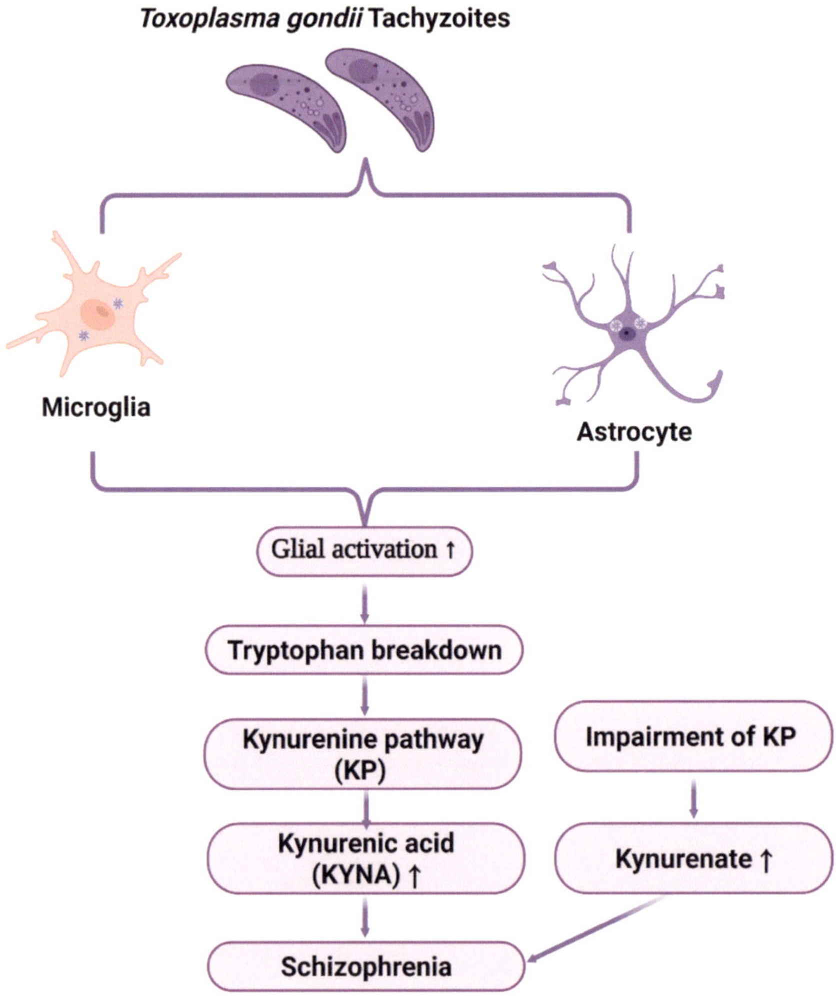

Figure 8.
T. gondii-mediated KYNA elevation and its association with schizophrenia.

In a broad sense, the parallels between the pathogenesis of these diseases are highly remarkable. It also shows their close relationship. These investigations show the molecular mechanism behind the pathogenesis of schizophrenia mediated by *T. gondii*.

7. Neurohistopathological findings of toxoplasmic encephalitis

Nonsuppurative and/or necrotizing meningoencephalitis are among the fundamental histological findings in TE. The existence of tissue cysts is a crucial observation. It has been shown that tissue cysts may be found throughout the brain, with the increased density in the cerebral cortex, hippocampus, thalamus, and amygdala regions being a noteworthy histopathological finding. The histological findings are neuronal degeneration and necrosis, localized gliosis foci, hyperemia,

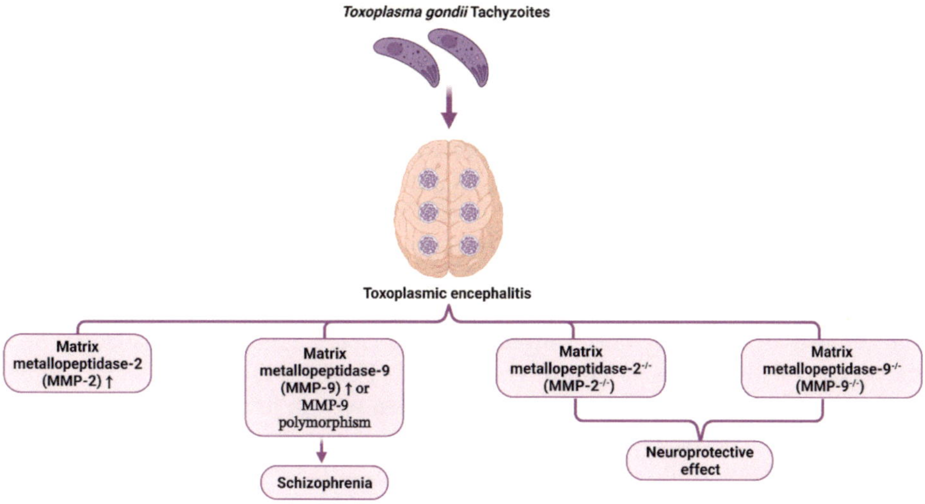

Figure 9.
Roles of MMP-2 and MMP-9 in TE neuropathology and schizophrenia.

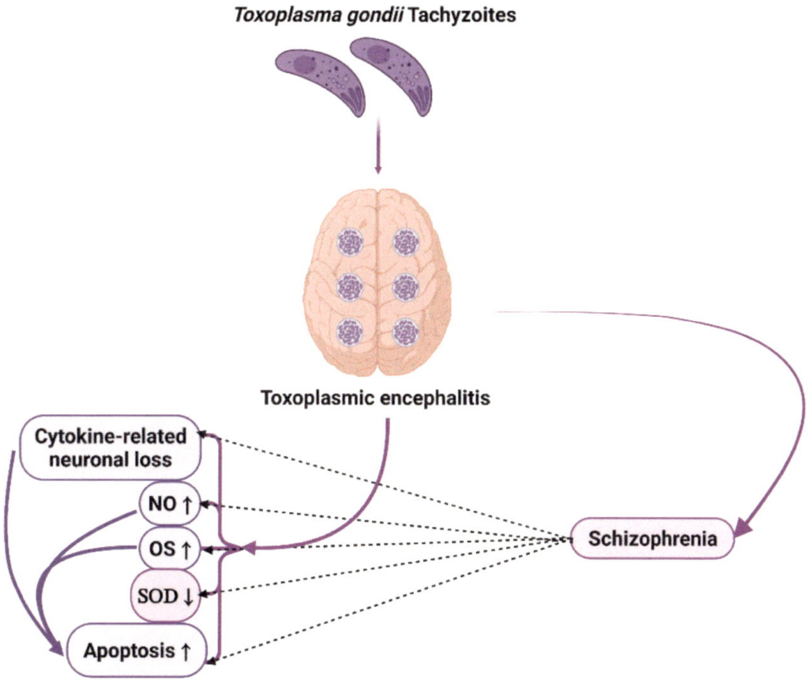

Figure 10.
The reason why TE continues to be a major health issue.

perivascular mononuclear cell infiltration, and capillary endothelial hypertrophy. As shown earlier, Purkinje cells are very susceptible to *T. gondii* tachyzoites. Histopathologically, these cells are also shrunken, with eosinophilic cytoplasm and necrotic [26–28, 86–88].

8. Diagnosis of toxoplasmic encephalitis

Multifocal necrotizing encephalitis is the main neuropathological sign of TE in in people living with AIDS [89, 90]. However, from the start of the epidemic until more recent years, a number of publications have used the term "toxoplasma abscess" [91]. Human Immunodeficiency Virus (HIV)-related TE does not exhibit pus, contrary to the traditional definition of a brain abscess, which requires for an intraparenchymal pus accumulation. The four histological phases of the development of a brain abscess are early cerebritis, late cerebritis, early capsule formation, and late capsule formation. Pus may be seen in the latter three stages [92]. Additionally, an immunocompromised host may have a considerable delay in the progression of the phases of a brain abscess [93]. Thus, early cerebritis that more closely resembles necrotizing encephalitis is often seen in toxoplasmosis patients [92]. Although these lesions are considered abscesses in some studies, we think it is more appropriate to define "toxoplasmic encephalitis" as researchers who have worked with TE for many years.

8.1 Diagnostic categories

There are important diagnostic criteria for TE. A differential diagnosis with diseases such as tuberculous meningitis or progressive multifocal leukoencephalopathy is required at this stage. In fact, the important points at this stage are early diagnosis and direct intervention, because it should never be forgotten that the time lost in neuropathology comes back with irreparable results [94].

8.1.1 Histopathology-confirmed toxoplasmic encephalitis

To establish the diagnosis clearly and definitively by histopathological examination, it is necessary to demonstrate the presence of *T. gondii* in brain biopsy or postmortem examination. Among the most commonly used methods, the priority is stereotactic computed tomography (CT)-guided needle biopsy. *T. gondii* tachyzoites or bradyzoites from tissue cysts can be seen at the margins of lesions by hematoxylin and eosin staining methods. However, the sensitivity can be significantly increased with immunohisto/cytochemical staining. Considering these histopathological methods, a definitive diagnosis can be made easily. However, the complications of brain biopsy should not be forgotten, and it should be done if necessary to make a definitive diagnosis. It should be recommended if there is no improvement in clinical findings after antitoxoplasma treatment and no progress is detected in imaging methods [94].

8.1.2 Laboratory-confirmed toxoplasmic encephalitis

For laboratory-confirmed TE, *T. gondii* DNA from cerebrospinal fluid (CSF) or brain biopsy material must be proven by nucleic acid amplification tests. The presence of serum *T. gondii* immunoglobulin G (IgG) antibodies also contributes significantly to diagnosing TE. Therefore, these two findings should be evaluated as a whole [94].

8.2 Important criteria for detecting suspected TE cases in immunocompromised patients

It has been extensively discussed above that immunocompromised patients are highly susceptible to TE, and serious pathology occurs in such cases. Here, we will focus on the steps to reveal a TE in these patients.

1. It should be considered when patients have clinical manifestations of massive brain lesions lasting up to 4 weeks, such as altered mental state, headache, or focal motor impairment [95].

2. Detection of the most common and diagnostic nodular lesions with perilesional edema and eccentric target sign in brain computed tomography (CT) scans is important. If any contradiction or negative result is seen at this stage, the patient should be referred to magnetic resonance imaging (MRI) without wasting time [96].

3. The clinical findings of patients receiving antitoxoplasma treatment with the suspicion of TE are followed closely, and a positive improvement is observed, and the improvement in neuropathology with imaging methods is an important advance for a definitive diagnosis.

4. The absence of *T. gondii* IgG antibodies and/or negative polymerase chain reaction (PCR) in blood samples/CSF taken from the patient does not completely exclude the possibility of TE. Instead, it requires redirection to other validation methods [97, 98].

Finally, it is worth mentioning that CNSToxoIndex evaluates the local synthesis of anti-toxoplasmic IgG in the CNS and their diffusion from blood as a consequence of blood-brain barrier disruption, and it is also a potent approach of TE diagnostics in HIV-positive individuals [99].

If a definitive diagnosis is desired, an important step in immunocompromised patients, lesions revealed by imaging methods should also be confirmed by histopathological and molecular methods.

9. Imaging features of toxoplasmic encephalitis

Imaging systems are of great importance in revealing the pathology caused by the diseases, both in the follow-up and diagnosis.

Since hematogenous distribution has a very important place in T. gongii infections, most of the lesions are seen in the basal ganglia, corticomedullary junction, white matter, and periventricular area [100]. The most accurate modality to diagnose TE is *via* magnetic resonance imaging (MR), which may also show the full scope of the illness. Infection of the brain produces lesions with poor attenuation on CT, low or high signal on T1WI (suggesting hemorrhagic components), and high signal on T2WI with surrounding edema. Currently, several important lesions in the imaging features of TE are very helpful in the diagnosis. Foremost among these, uniform nodular or multifocal ring-enhancing lesions are seen, particularly in the cortex and periventricular white matter [100, 101]. Lesions from toxoplasmosis exhibit uniform nodular or ring enhancement [100, 101]. If there is poor enhancement, it may be modest or nonexistent in immunocompromised individuals depending on their cellular immune response. The "concentric target sign" in T2-weighted imaging and the "eccentric target sign" in postcontrast T1-weighted sequences are significant MRI findings reported in TE [100, 102–105]. In particular, this finding is considered pathognomonic. Multiple lesions were seen in the bilateral basal ganglia, frontal, temporal, parietal, occipital, and cerebellar lobes, as well as ring enhancement on postcontrast sequence near edema and the development of "eccentric target sign" on an MRI of the brain with contrast. These lesions featured intra-lesional hemorrhagic foci and alternating hyper and hypointense zones on T2-weighted sequence, along with perilesional edema that was clearly visible and corresponded to the "concentric target sign" [105].

Susceptibility-weighted imaging (SWI) of hemorrhagic lesions commonly seen in TE was superior and more sensitive than GRE T2*WI. It is emphasized that diffuse hemorrhage foci seen in TE and frequently detected in recent studies may also be because of the superior sensitivity of SWI in detecting hemorrhage [106].

The eccentric target sign linked histologically to a sizable necrotizing abscess. The enhancing rim histologically corresponded to a thick ring of histiocytic response with inflamed and proliferating vessels demarcated by immunostaining to Factor VIII related antigen of vascular endothelium, while the surrounding zone of necrosis caused the intermediate zone of hypointensity. On T2W images, the perilesional white matter edema with demyelination appeared as strong hyperintensities [107]. The ring-enhancing lesion's periphery occasionally contained tissue cysts containing *T. gondii* bradyzoite, while the necrotic zone's inflamed vascular zone contained diffusely dispersed tachyzoite forms that could be seen by immunohistochemistry using antibodies to tachyzoite-specific antigen (p30). Eccentric target sign, a characteristic of imaging, is thought to be strongly predictive of toxoplasmosis. Less than 30% of cases have this sign, which may be seen on postcontrast CT scans (asymmetric target sign) or T1W MRI scans [108]. There are three alternating zones in this: a centrally eccentric innermost enhancing core, a middle hypointense zone, and an outermost hyperintense enhancing rim. A leash of inflamed, leaking arteries that enter the lesion *via* a sulcus create the eccentric core [107].

Diffusion- and perfusion-weighted imaging (DWI and PWI) and spectroscopy may help distinguish toxoplasmosis from other disorders that mirror it, such as brain lymphoma and infections, especially in immunocompromised individuals. Toxoplasmosis may exhibit peripheral limited diffusion, but pyogenic abscesses often exhibit central restricted diffusion. PWI in toxoplasmosis is minimal, but high in

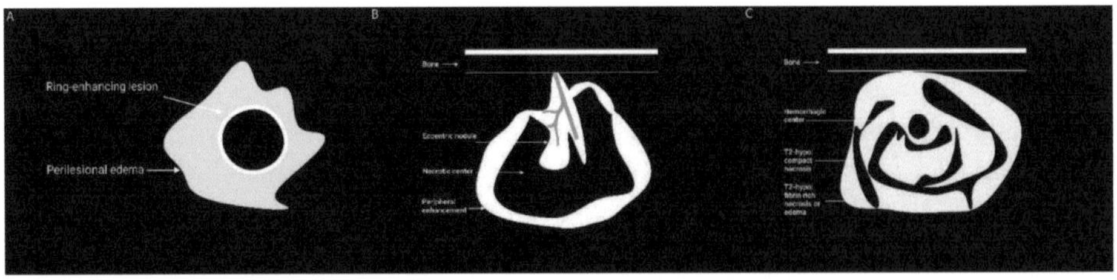

Figure 11.
Ring-enhancing lesions surrounding hyperintensity, consistent with vasogenic edema, eccentric target sign, and concentric target sign.

lymphoma and other neoplasms. Toxoplasma and pyogenic abscesses both had an increased lipid-lactate peak in spectroscopy [109, 110]. PET-CT and thallium single-photon emission computed tomography often reveal increased uptake in lymphoma and decreased uptake in toxoplasmosis (**Figure 11**) [109, 111].

10. Conclusion

The mechanisms of neuroimmunopathogenesis, the most significant and complicated aspect of TE, were covered in this chapter, which was the product of much study and effort. Each part has been thoroughly evaluated and clarified. Therefore, the most fundamental aspect of avoiding neuropathology induced by TE, which may lead to mortality, is the investigation and complete disclosure of these processes. This book chapter, which discusses this extensive neuroimmunopathogenesis and will play an essential role in developing novel therapeutic medications, will make a significant addition to the body of literature. Furthermore, diagnostic methods that are critical are emphasized. The tricks of imaging methods are summarized, and key points are highlighted. It is hoped that considering all aspects of TE, which has a complex pathological mechanism, will make a significant contribution to the literature.

Conflict of interest

The authors declare that there is no conflict of interests regarding the publication of this chapter.

Author details

Gungor Cagdas Dincel[1], Abdulaziz Alouffi[2], Ebtesam M. Al Olayan[3], Guillermo Tellez-Isaias[4] and Saeed El-Ashram[5,6]*

1 Eskil Vocational School, Laboratory and Veterinary Science, Aksaray University, Aksaray, Turkey

2 King Abdulaziz City for Science and Technology, Riyadh, Saudi Arabia

3 Department of Zoology, Faculty of Science, King Saud University, Riyadh, Saudi Arabia

4 Department of Poultry Science, University of Arkansas, Fayetteville, AR, USA

5 Faculty of Science, Kafrelsheikh University, Kafr El-Sheikh, Egypt

6 College of Life Science and Engineering, Foshan University, Foshan, China

*Address all correspondence to: saeed_elashram@yahoo.com

IntechOpen

© 2023 The Author(s). Licensee IntechOpen. This chapter is distributed under the terms of the Creative Commons Attribution License (http://creativecommons.org/licenses/by/3.0), which permits unrestricted use, distribution, and reproduction in any medium, provided the original work is properly cited.

References

[1] Hill DE, Chirukandoth S, Dubey JP. Biology and epidemiology of *Toxoplasma gondii* in man and animals. Animal Health Research Reviews. 2005;**6**:41-61

[2] Dubey JP. Toxoplasmosis of Animals and Humans. 2second ed. Boca Raton, Florida: CRC Press; 2009. pp. 1-71

[3] Kasper LH, Boothroyd JC. *Toxoplasma gondii*: Immunology and molecular biology. In: Warren K, Agabian N, editors. Immunology of Parasitic Infections. UK: Oxford; 1992. pp. 269-301

[4] Bhopale GM. Pathogenesis of toxoplasmosis. Comparative Immunology, Microbiology and Infectious Diseases. 2003;**26**:213-222

[5] Hamdani N, Daban-Huard C, Lajnef M, Richard JR, Delavest M, Godin O, et al. Relationship between *Toxoplasma gondii* infection and bipolar disorder in a French sample. Journal of Affective Disorders. 2013;**148**:444-448

[6] Torrey EF, Yolken RH. *Toxoplasma gondii* and schizophrenia. Emerging Infectious Diseases. 2003;**9**:1375-1380

[7] Mortensen PB, Nørgaard-Pedersen B, Waltoft BL, Sørensen TL, Hougaard D, Yolken RH. Early infections of *Toxoplasma gondii* and the later development of schizophrenia. Schizophrenia Bulletin. 2007;**33**:741-744

[8] Montoya JG, Liesenfeld O. Toxoplasmosis. Lancet. 2004;**363**(9425):1965-1976

[9] Machala L, Kodym P, Maly M, Geleneky M, Beran O, Jilich D. Toxoplasmosis in immunocompromised patients. Epidemiologie, Mikrobiologie, Imunologie. 2015;**64**:59-65

[10] Frenkel JK, Amare M, Larsen W. Immune competence in a patient with Hodgkin's disease and relapsing toxoplasmosis. Infection. 1978;**6**:84-91

[11] Maciel E, Siqueira I, Queiroz AC, Melo A. *Toxoplasma gondii* myelitis in a patient with adult T-cell leukemia-lymphoma. Arquivos de Neuro-Psiquiatria. 2000;**58**:1107-1109

[12] Kojima M, Nakamura N, Murayama K, Igarashi T, Matsumoto M, Matsuda H, et al. Reactive lymphoid hyperplasia with giant follicles associated with a posttherapeutic state of hematological malignancies. A report of eight cases. Tumori. 2010;**96**:143-148

[13] Chehrazi-Raffle A, Luu M, Yu Z, Liou F, Kittleson M, Hamilton M, et al. *Toxoplasma gondii* serology and outcomes after heart transplantation: Contention in the literature. Transplantation Proceedings. 2015;**47**:1949-1953

[14] Luft BJ, Hafner R, Korzun AH, Leport C, Antoniskis D, Bosler EM, et al. Toxoplasmic encephalitis in patients with the acquired immunodeficiency syndrome. Members of the ACTG 077p/ANRS 009 Study Team. New England Journal of Medicine. 1993;**329**:995-1000

[15] Jones JL, Hanson DL, Chu SY, Ciesielski CA, Kaplan JE, Ward JW, et al. Toxoplasmic encephalitis in HIV-infected persons: Risk factors and trends. The Adult/Adolescent Spectrum of Disease Group. AIDS. 1996;**10**:1393-1399

[16] Walker M, Zunt JR. Parasitic central nervous system infections in immunocompromised hosts. Clinical Infectious Diseases. 2005;**40**:1005-1015

[17] Abbott NJ, Ronnback L, Hansson E. Astrocyte-endothelial interactions at the blood-brain barrier. Nature Reviews. Neuroscience. 2006;7:41-53

[18] Abbott NJ, Patabendige AA, Dolman DE, Yusof SR, Begley DJ. Structure and function of the blood-brain barrier. Neurobiology of Disease. 2010;**37**:13-25

[19] Nakagawa S, Deli MA, Kawaguchi H, Shimizudani T, Shimono T, Kittel A, et al. A new blood-brain barrier model using primary rat brain endothelial cells, pericytes and astrocytes. Neurochemistry International. 2009;**54**:253-263

[20] Deli MA, Abraham CS, Kataoka Y, Niwa M. Permeability studies on in vitro blood-brain barrier models: Physiology, pathology, and pharmacology. Cellular and Molecular Neurobiology. 2005;**25**:59-127

[21] Olivera GC, Ross EC, Peuckert C, Barragan A. Blood-brain barrier-restricted translocation of Toxoplasma gondii from cortical capillaries. eLife. 2021;**8**(10):e69182

[22] Lachenmaier SM, Deli MA, Meissner M, Liesenfeld O. Intracellular transport of *Toxoplasma gondii* through the blood-brain barrier. Journal of Neuroimmunology. 2011;**232**:119-130

[23] Masocha W, Kristensson K. Passage of parasites across the bloodbrain barrier. Virulence. 2012;**3**:202-212

[24] Konradt C, Ueno N, Christian DA, Delong JH, Pritchard GH, Herz J, et al. Endothelial cells are a replicative niche for entry of *Toxoplasma gondii* to the central nervous system. Nature Microbiology. 2016;**15**(1):16001

[25] Delorme-Walker V, Abrivard M, Lagal V, Anderson K, Perazzi A, Gonzalez V, et al. Toxofilin upregulates the host cortical actin cytoskeleton dynamics facilitating Toxoplasma invasion. Journal of Cell Science. 2012;**125**:4333-4342

[26] Dincel GC, Atmaca HT. Increased expressions of ADAMTS-13 and apoptosis contribute to neuropathology during *Toxoplasma gondii* encephalitis in mice. Neuropathology. 2016;**36**(3):211-226

[27] Dincel GC, Atmaca HT. Role of oxidative stress in the pathophysiology of *Toxoplasma gondii* infection. International Journal of Immunopathology and Pharmacology. 2016;**29**(2):226-240

[28] Dincel GC, Atmaca HT. Nitric oxide production increases during *Toxoplasma gondii* encephalitis in mice. Experimental Parasitology. 2015;**156**:104-112

[29] Noubade R, delRio R, Mcelvany B, Millward JM, Wagner DD, Offner H, et al. von-Willebrand factor influences blood brain barrier permeability and brain inflammation in experimental allergic encephalomyelitis. The American Journal of Pathology. 2008;**173**(3):892-900

[30] Dincel GC, Kul O. Increased expressions of ADAMTS-13, neuronal nitric oxide synthase, and neurofilament correlate with severity of neuropathology in Border disease virus-infected small ruminants. PLoS One. 2015;**23**(3):e0120005

[31] Karayigit MO, Dincel GC. Role of ADAMTS-13 and nNOS expression in neuropathogenesis of listeric encephalitis of small ruminants. Biotechnic & Histochemistry. 2020;**95**(8):584-596

[32] Benevides L, Milanezi CM, Yamauchi LM, Benjamim CF, Silva JS,

Silva NM. CCR2 receptor is essential to activate microbicidal mechanisms to control *Toxoplasma gondii* infection in the central nervous system. The American Journal of Pathology. 2008;**173**(3):741-751

[33] Batista SJ, Still KM, Johanson D, Thompson JA, Carleigh A. Harris Gasdermin-D-dependent IL-1α release from microglia promotes protective immunity during chronic *Toxoplasma gondii* infection. Nature. Communications. 2020;**11**:3687

[34] Schlüter D, Deckert-Schlüter M, Lorenz E, Meyer T, Röllinghoff M, Bogdan C. Inhibition of inducible nitric oxide synthase exacerbates chronic cerebral toxoplasmosis in *Toxoplasma gondii*-susceptible C57BL/6 mice but does not reactivate the latent disease in *T. gondii*-resistant BALB/c mice. Journal of Immunology. 1999;**162**:3512-3518

[35] Hayashi S, Chan CC, Gazzinelli R, Roberge FG. Contribution of nitric oxide to the host-parasite equilibrium in toxoplasmosis. Journal of Immunology. 1996;**156**(4):1476-1481

[36] Scharton-Kersten TM, Yap G, Magram J, Sher A. Inducible nitric oxide is essential for host control of persistent but not acute infection with the intracellular pathogen *Toxoplasma gondii*. The Journal of Experimental Medicine. 1997;**185**:1261-1273

[37] Gazzinelli RT, Eltoum I, Wynn TA, Sher A. Acute cerebral toxoplasmosis is induced by in vivo neutralization and correlates with the down-regulated expression of inducible nitric oxide synthase and other markers of macrophage activation. Journal of Immunology. 1993;**151**:3672-3681

[38] Khan IA, Schwartzman JD, Matsuura T, Kasper LH. A dichotomous role of nitric oxide during acute *Toxoplasma gondii* infection in mice. Proceedings of the National Academy of Sciences of the United States of America. 1997;**94**(25):13955-13960

[39] Sibley LD, Adams LB, Fukutomi Y, Krahenbuhl JL. Tumor necrosis factor alpha triggers antitoxoplasmal activity of IFN-gamma primed macrophages. Journal of Immunology. 1991;**147**:2340-2345

[40] Candolfi E, Hunter CA, Remington JS. Mitogen- and antigen specific proliferation of T cells in murine toxoplasmosis is inhibited by reactive nitrogen intermediates. Infection and Immunity. 1994;**62**(5):1995-2001

[41] Bohne W, Heesemann J, Gross U. Reduced replication of *Toxoplasma gondii* is necessary for induction of bradyzoite-specific antigens: A possible role for nitric oxide in triggering stage conversion. Infection and Immunity. 1994;**62**:1761-1767

[42] Ibrahim HM, Bannai H, Xuan X, Nishikawa Y. *Toxoplasma gondii* cyclophilin 18-mediated production of nitric oxide induces bradyzoite conversion in a CCR5-dependent manner. Infection and Immunity. 2009;**77**(9):3686-3695

[43] Peterson PK, Gekker G, Hu S, Chao CC. Human astrocytes inhibit intracellular multiplication of *Toxoplasma gondii* by a nitric oxide-mediated mechanism. The Journal of Infectious Diseases. 1995;**171**:516-518

[44] Al-Chalabi A, Miller CC. Neurofilaments and neurological disease. BioEssays. 2003;**25**(4):346-355

[45] Liu Q, Xie F, Alvarado-Diaz A, Smith MA, Moreira PI, Zhu X, et al. Neurofilamentopathy in

neurodegenerative diseases. Open Neurology Journal. 2011;**5**:58-62

[46] Norgren N, Rosengren L, Stigbrand T. Elevated neurofilament levels in neurological diseases. Brain Research. 2003;**987**:25-31

[47] Hatfield RH, McKernan RM. CSF neuronspecific enolase as a quantitative marker of neuronal damage in a rat stroke model. Brain Research. 1992;**577**:249-252

[48] Thangavel R, Stolmeier D, Yang X, Zaheer A. Expression of glia maturation factor in neuropathological lesions of Alzheimer's disease. Neuropathology and Applied Neurobiology. 2012;**38**:572-581

[49] Zaheer A, Zaheer S, Sahu SK, Yang B, Lim R. Reduced severity of experimental autoimmune encephalomyelitis in GMF-deficient mice. Neurochemical Research. 2007;**32**:39-47

[50] Zaheer A, Zaheer S, Sahu SK, et al. A novel role of glia maturation factor: Induction of granulocyte-macrophage colony-stimulating factor and pro-inflammatory cytokines. Journal of Neurochemistry. 2007;**101**:364-376

[51] Dincel GC, Kul O. First description of enhanced expression of transforming growth factor-alpha (TGF-α) and glia maturation factor-beta (GMF-β) correlate with severity of neuropathology in border disease virus-infected small ruminants. Microbial Pathogenesis. 2019;**128**:301-310

[52] Thangavel R, Kempuraj D, Stolmeier D, Anantharam P, Khan M, Zaheer A. Glia maturation factor expression in entorhinal cortex of Alzheimer's disease brain. Neurochemical Research. 2013;**38**:1777-1784

[53] Dincel GC. First description of enhanced expression of glia maturation factor-beta in experimental toxoplasmic encephalitis. The Journal of International Medical Research. 2017;**45**(6):1670-1679

[54] An R, Tang Y, Chen L, Cai H, Lai DH, Liu K, et al. Encephalitis is mediated by ROP18 of *Toxoplasma gondii*, a severe pathogen in AIDS patients. Proceedings of the National Academy of Sciences of the United States of America. 2018;**115**(23):5344-5352

[55] Halonen SK, Taylor GA, Weiss LM. Gamma interferon-induced inhibition of *Toxoplasma gondii* in astrocytes is mediated by IGTP. Infection and Immunity. 2001;**69**:5573-5576

[56] Hidano S, Randall LM, Dawson L, Dietrich HK, Konradt C, Klover PJ, et al. STAT1 signaling in astrocytes is essential for control of infection in the central nervous system. MBio. 2016;7:e01881-e01816

[57] Beiting DP, Hidano S, Baggs JE, Geskes JM, Fang Q, Wherry EJ, et al. The orphan nuclear receptor TLX is an enhancer of STAT1-mediated transcription and immunity to *Toxoplasma gondii*. PLoS Biology. 2015;**13**(7):e1002200

[58] Drögemüller K, Helmuth U, Brunn A, Sakowicz-Burkiewicz M, Gutmann DH, Mueller W, et al. Astrocyte gp130 expression is critical for the control of *Toxoplasma encephalitis*. Journal of Immunology. 2008;**181**:2683-2693

[59] Fagard R, Van Tan H, Creuzet C, Pelloux H. Differential development of *Toxoplasma gondii* in neural cells. Parasitology Today. 1999;**15**(12):504-507

[60] Fischer HG, Nitzgen B, Reichmann G, Hadding U. Cytokine responses induced by *Toxoplasma gondii* in astrocytes and microglial cells.

European Journal of Immunology. 1997;**27**:1539-1548

[61] Bjerkas I. Neuropathology and host-parasite relationship of acute experimental toxoplasmosis of the blue fox (alopexlagopus). Veterinary Pathology. 1990;**27**:381-390

[62] Gonzalez LE, Rojnik B, Urrea F, Urdaneta H, Petrosino P, Colasante C, et al. *Toxoplasma gondii* infection lower anxietyas measured in the plus-maze and social interaction tests in rats A behavioral analysis. Behavioural Brain Research. 2007;**177**:70-79

[63] Wilson EH, Hunter CA. The role of astrocytes in the immunopathogenesis of toxoplasmic encephalitis. International Journal for Parasitology. 2004;**34**:543-548

[64] Stenzel W, Soltek S, Schlüter D, Deckert M. The intermediate filament GFAP is important for the control of experimental murine *Staphylococcus aureus*-induced brain abscess and *Toxoplasma encephalitis*. Journal of Neuropathology and Experimental Neurology. 2004;**63**(6):631-640

[65] Cekanaviciute E, Dietrich HK, Axtell RC, Williams AM, Egusquiza R, Wai KM, et al. Astrocytic TGF-β signaling limits inflammation and reduces neuronal damage during central nervous system Toxoplasma infection. Journal of Immunology. 2014;**193**(1):139-149

[66] Rozenfeld C, Martinez R, Figueiredo RT, Bozza MT, Lima FR, Pires AL, et al. Soluble factors released by *Toxoplasma gondii*-infected astrocytes down-modulate nitric oxide production by gamma interferon-activated microglia and prevent neuronal degeneration. Infection and Immunity. 2003;**71**:2047-2057

[67] Espey MG, Chernyshev ON, Reinhard JF Jr, Namboodiri MA, Colton CA. Activated human microglia produce the excitotoxin quinolinic acid. Neuroreport. 1997;**8**(2):431-434

[68] Guillemin GJ, Kerr SJ, Smythe GA, Smith DG, Kapoor V, Armati PJ, et al. Kynurenine pathway metabolism in human astrocytes: A paradox for neuronal protection. Journal of Neurochemistry. 2001;**78**(4):842-853

[69] Heyes MP, Achim CL, Wiley CA, Major EO, Saito K, Markey SP. Human microglia convert l-tryptophan into the neurotoxin quinolinic acid. The Biochemical Journal. 1996;**320**(2):595-597

[70] Schwarcz R, Rassoulpour A, Wu HQ, Medoff D, Tamminga CA, Roberts RC. Increased cortical kynurenate content in schizophrenia. Biological Psychiatry. 2001;**50**(7):521-530

[71] Notarangelo FM, Wilson EH, Horning KJ, Thomas MAR, Harris TH, Fang Q, et al. Evaluation of kynurenine pathway metabolism in *Toxoplasma gondii*-infected mice: Implications for schizophrenia. Schizophrenia Research. 2014;**152**(1):261-267

[72] El Mouhawess A, Hammoud A, Zoghbi M, et al. Relationship between *Toxoplasma gondii* seropositivity and schizophrenia in the Lebanese population: Potential implication of genetic polymorphism of MMP-9. BMC Psychiatry. 2020;**20**:264

[73] Lepeta K, Kaczmarek L. Matrix Metalloproteinase-9 as a novel player in synaptic plasticity and schizophrenia. Schizophrenia Bulletin. 2015;**41**(5):1003-1009

[74] Rybakowski JK. Matrix Metalloproteinase-9 (MMP9):

A mediating enzyme in cardiovascular disease, cancer, and neuropsychiatric disorders. Cardiovascular Psychiatry and Neurology. 2009;**2009**:904836

[75] Lu C-Y, Lai S-C. Induction of matrix metalloproteinase-2 and -9 via Erk1/2-NF-κB pathway in human astroglia infected with toxoplasma gondii. Acta Tropica. 2013;**127**(1):14-20

[76] Lu C-Y, Lai S-C. Matrix metalloproteinase-2 and -9 lead to fibronectin degradation in astroglia infected with toxoplasma gondii. Acta Tropica. 2013;**125**(3):320-329

[77] Nasyrova RF, Ivashchenko DV, Ivanov MV, Neznanov NG. Role of nitric oxide and related molecules in schizophrenia pathogenesis: Biochemical, genetic and clinical aspects. Frontiers in Physiology. 2015;**6**:139

[78] Ermakov EA, Dmitrieva EM, Parshukova DA, Kazantseva DV, Vasilieva AR, Smirnova LP. Oxidative stress-related mechanisms in schizophrenia pathogenesis and new treatment perspectives. Oxidative Medicine and Cellular Longevity. 2021;**2021**:8881770

[79] Bošković M, Vovk T, Kores Plesničar B, Grabnar I. Oxidative stress in schizophrenia. Current Neuropharmacology. 2011;**9**(2):301-312

[80] Mukerjee S, Mahadik SP, Scheffer R, Correnti EE, Kelkar H. Impaired antioxidant defense at the onset of psychosis. Schizophrenia Research. 1996;**19**(1):19-26

[81] Jarskog LF, Gilmore JH, Selinger ES, Lieberman JA. Cortical bcl-2 protein expression and apoptotic regulation in schizophrenia. Biological Psychiatry. 2000;**48**(7):641-650

[82] Catts VS, Catts SV. Apoptosis and schizophrenia: Is the tumour suppressor gene, p53, a candidate susceptibility gene? Schizophrenia Research. 2000;**41**(3):405-415

[83] Jarskog LF, Glantz LA, Gilmore JH, Lieberman JA. Apoptotic mechanisms in the pathophysiology of schizophrenia. Progress in Neuro-Psychopharmacology & Biological Psychiatry. 2005;**29**(5):846-858

[84] Glantz LA, Gilmore JH, Lieberman JA, Jarskog LF. Apoptotic mechanisms and the synaptic pathology of schizophrenia. Schizophrenia Research. 2006;**81**(1):47-63

[85] Marx CE, Jarskog LF, Lauder JM, Lieberman JA, Gilmore JH. Cytokine effects on cortical neuron MAP-2 immunoreactivity: Implications for schizophrenia. Biological Psychiatry. 2001;**50**(10):743-749

[86] El Saftawy EA, Shash RY, Aboulhoda BE, Arsanyos SF, Albadawi EA, Abou-Fandoud SM, et al. Insights into immunopathology and triggering of apoptosis in chronic cerebral toxoplasmosis using murine models. Tropical Biomedicine. 2021;**38**(2):53-62

[87] Melzer TC, Cranston HJ, Weiss LM, Halonen SK. Host cell preference of toxoplasma gondii cysts in murine brain: A Confocal Study. Journal of neuroparasitology. 2010;**1**:N100505

[88] Haroon F, Händel U, Angenstein F, Goldschmidt J, Kreutzmann P, Lison H, et al. *Toxoplasma gondii* actively inhibits neuronal function in chronically infected mice. PLoS One. 2012;**7**(4):e35516

[89] Levy RM, Bredesen DE, Rosenblum ML. Opportunistic central nervous system pathology in patients

with AIDS. Annals of Neurology. 1988;**23**:S7-S12

[90] Sher JH. Cerebral toxoplasmosis. Lancet. 1983;**1**(8335):1225

[91] Brouwer MC, Tunkel AR, McKhann GM 2nd, van de Beek D. Brain abscess. The New England Journal of Medicine. 2014;**371**(5):447-456

[92] Britt RH, Enzmann DR. Clinical stages of human brain abscesses on serial CT scans after contrast infusion. Computerized tomographic, neuropathological, and clinical correlations. Journal of Neurosurgery. 1983;**59**(6):972-989

[93] Obana WG, Britt RH, Placone RC, Stuart JS, Enzmann DR. Experimental brain abscess development in the chronically immunosuppressed host. Computerized tomographic and neuropathological correlations. Journal of Neurosurgery. 1986;**65**(3):382-391

[94] Vidal JE. HIV-related cerebral toxoplasmosis revisited: Current concepts and controversies of an old disease. Journal of International Association. 2019;**2019**:182

[95] Lyons RE, McLeod R, Roberts CW. *Toxoplasma gondii* tachyzoite-bradyzoite interconversion. Trends in Parasitology. 2002;**18**(5):198-201

[96] Lee SB, Lee TG. Toxoplasmic Encephalitis in patient with acquired immunodeficiency syndrome. Brain Tumor Research Treatment. 2017;**5**(1):34-36

[97] Vilaseca J, Arnau JM, Bacardi R, Mieras C, Serrano A, et al. Kaposi's sarcoma and *Toxoplasma gondii* brain abscess in a Spanish homosexual. Lancet. 1982;**1**(8271):572

[98] Luft BJ, Conley F, Remington JS, Laverdiere M, Wagner KF, et al. Outbreak of central-nervous-system toxoplasmosis in western Europe and North America. Lancet. 1983;**1**(8328):781-784

[99] Bondarenko AV, Katsapov DV, Gavrylov AV, Didova TV, Nahornyi IA. Immunodiagnostics of cerebral toxoplasmosis depending on permeability of blood-brain barrier. Wiadomości Lekarskie. 2020;**73**(2):285-288

[100] Mueller-Mang C, Mang TG, Kalhs P, Thurnher MM. Imaging characteristics of toxoplasmosis encephalitis after bone marrow transplantation: Report of two cases and review of the literature. Neuroradiology. 2006;**48**(2):84-89

[101] Smith A, Smirniotopoulos J, Rushing E. Central nervous system infections associated with human immunodeficiency virus infection: Radiologic-pathologic Correlation1. Radiographics. 2008:282033-282058

[102] Roche AD, Rowley D, Brett FM, Looby S. Concentric and eccentric target MRI signs in a case of HIV-associated cerebral toxoplasmosis. Case Report Neurological Medicine. 2018;**2018**:9876514

[103] Mahadevan A, Ramalingaiah AH, Parthasarathy S, Nath A, Ranga U, et al. Neuropathological correlate of the "concentric target sign" in MRI of HIV-associated cerebral toxoplasmosis. Journal of Magnetic Resonance Imaging. 2013;**38**(2):488-495

[104] Bansal S, Goyal M, Modi M, Ahuja C, Lal V. Eccentric target sign of cerebral toxoplasmosis. QJM. 2016;**109**(8):555

[105] Masamed R, Meleis A, Lee EW, Hathout GM. Cerebral toxoplasmosis:

Case review and description of a new imaging sign. Clinical Radiology. 2009;**64**:560-563

[106] Benson JC, Cervantes G, Baron TR, Tyan AE, Flanagan S, Lucato LT, et al. Imaging features of neurotoxoplasmosis: A multiparametric approach, with emphasis on susceptibility-weighted imaging. European Journalof Radiological Open. 2018;**5**:45-51

[107] Kumar GG, Mahadevan A, Guruprasad AS, Kovoor JM, Satishchandra P, et al. Eccentric target sign in cerebral toxoplasmosis: Neuropathological correlate to the imaging feature. Journal of Magnetic Resonance Imaging. 2010;**31**(6):1469-1472

[108] Ramsey RG, Geremia GK. CNS complications of AIDS: CT and MR findings. AJR. American Journal of Roentgenology. 1988;**151**(3):449-454

[109] Muccio CF, Caranci F, D'Arco F, Cerase A, De Lipsis L, et al. Magnetic resonance features of pyogenic brain abscesses and differential diagnosis using morphological and functional imaging studies: A pictorial essay. Journal of Neuroradiology. 2014;**41**(3):153-167

[110] Xu XX, Li B, Yang HF, Du Y, Li Y, et al. Can diffusion-weighted imaging be used to differentiate brain abscess from other ring-enhancing brain lesions? A meta-analysis. Clinical Radiology. 2014;**69**(9):909-915

[111] Hussain FS, Hussain NS. Clinical utility of Thallium-201 single photon emission computed tomography and cerebrospinal fluid Epstein-Barr virus detection using polymerase chain reaction in the diagnosis of AIDS-related primary central nervous system lymphoma. Cureus. 2016;**8**(5):e606

Chapter 6

The Important Role of Laboratories in the Diagnosis and Prevention of Toxoplasma Infection

María Vanessa Schneider

Abstract

Toxoplasma gondii is a high universal distribution intracellular parasite that causes toxoplasmosis, an infection of significant importance when pregnant women or immunosuppressed patients become infected. In immunocompetent humans, it is rarely a symptomatic disease. Diagnosis and prevention are essential to reduce severe health consequences in susceptible patients. Laboratories must play an important role in trying to avoid seroconversion, by helping in prevention and accurate diagnosis to facilitate patients' access to treatment as early as possible. The main focus of this chapter is about the steps laboratories should take in diagnosing *Toxoplasma gondii* infection by using screening tests (ELISA, HAI, AD) or confirmatory tests (Sabin Feldman dye test, ISAGA, PCR, Isolation)—determining whether patients are toxoplasmosis positive or negative—and in being actively involved in the prevention of the infection. Failing to do this would make the future of these families a very difficult road to travel.

Keywords: toxoplasmosis, *toxoplasma gondii* infection, seroconversion, congenital toxoplasmosis, laboratory diagnosis, ELISA, Sabin Feldman dye test, ISAGA, PCR, isolation

1. Introduction

Toxoplasma gondii infection affects a third of the world's population. Although in people with a healthy immune system it does not cause disease and it is usually asymptomatic, in pregnancy and immunosuppressed patients it can cause serious side effects.

In pregnancy, congenital toxoplasmosis may cause severe consequences to the fetus, from death in uterus to late important clinical conditions such as cognitive impairment, blurry vision by inflammation of the retina, deafness, epilepsy, seizures, tuberculosis, *Pneumocystis jiroveci* pneumonia, or poor coordination.

In primary or secondary immunosuppressed patients who get *Toxoplasma gondii* infection, or a reactivation of the disease, unwanted sequelae can also occur [1–4].

Laboratories should play an important role. "The important role of laboratories" should be getting involved and assuming the commitment not only in the diagnosis, but also in the prevention. They must bridge the gap between the attending physician and patients.

IntechOpen

1.1 Pregnant women testing

If toxoplasmosis serology is unknown before women became pregnant, then start by finding it out as early as possible within the first trimester [1–4].

An IgG test is the first thing to do to know the patient's serology. Which test should be selected? Laboratories must maximize their available equipment, resources, and capabilities, to choose their tests. Do any test, it is better than doing nothing.

Once an IgG result is obtained, two paths will open:

1. If IgG serology result is negative, great care must be taken because even though these women are not infected, they are susceptible to become infected [3–6]. This is why they must be followed. How? By warning about primary prevention measures (see Preventive Medicine) [1, 3, 4, 7, 8] and monitoring their IgG serology every month or at least once every three months to make sure not become infected during the rest of pregnancy (as shown in **Figure 1a**) [2, 9, 10].

2. If IgG serology result is positive, then the next step is to do a Toxo IgM test to be sure that it is not a potential seroconversion [1–4].

1.2 Toxo IgM test result

After the IgM test result is ready, two new situations emerge the following:

1. If IgM result is negative means these women were toxoplasmosis positive before pregnancy. In this case, it is a past infection and they are immunized against the parasite infection. There is no risk of becoming infected during their pregnancy and their children are out of danger (displayed in **Figure 1b**)

2. If IgM result is positive, there is a high suspicion of a seroconversion (SC), in which case this must be confirmed (as illustrated in **Figure 1c**).

1.3 How to confirm a seroconversion

The way to confirm a SC will depend on capabilities and the laboratory scope. Now at this point there are two other new scenarios:

1. Routine Laboratories: could refer this sample to a reference laboratory to perform more specific tests such as Western blot, PCR, ISAGA M, ISAGA A, ISAGA E, and Avidity test [2]. If there is not a reference laboratory or a more complex one nearby, then deem the result equivocal with an observation suggesting to repeat the test in three weeks. When the new sample arrives and an increase in title is observed, seroconversion is confirmed [1–5, 7, 8, 11].

2. Reference Laboratories: must enlarge the sample with other specific tests such as those named above. At the time of delivery, isolation of *Toxoplasma gondii* parasite in placenta and umbilical blood cord can be done and, furthermore, studied the type of strain to which it belongs to know the degree of impact to the newborn [12–17].

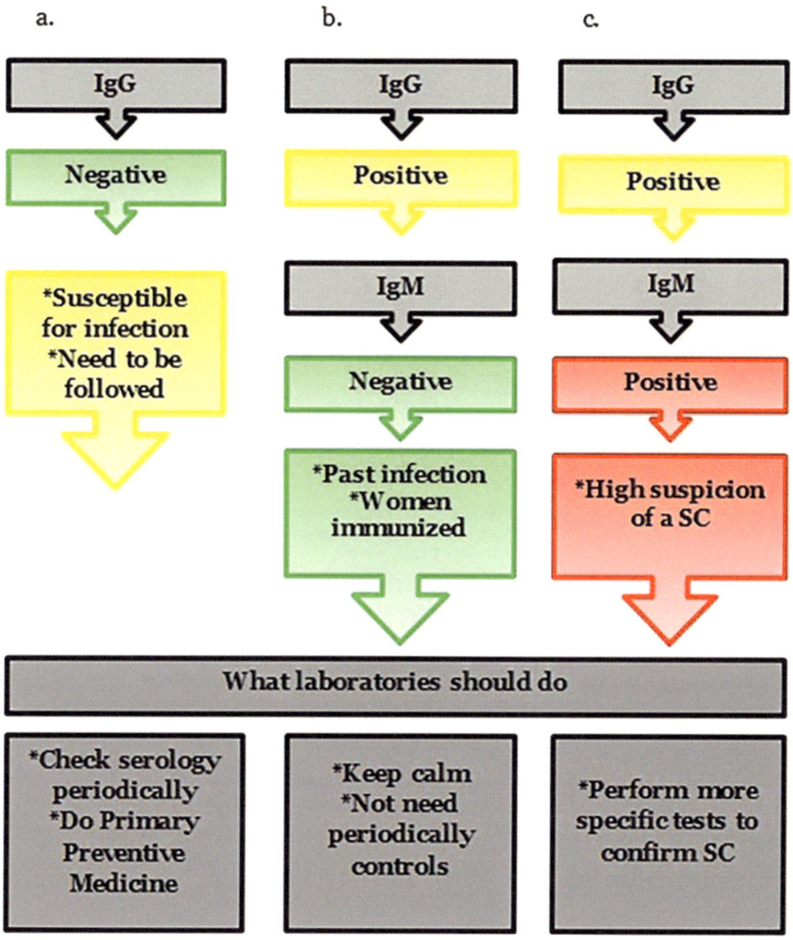

Figure 1.
Pregnant women-testing. a. Negative. b. Past infection. c. Seroconversion. Note: Own elaboration.

Once SC is confirmed, the doctor in charge must be notified as soon as possible, so these women can be medicated to avoid parasites infect placenta and make sure the newborn does not have any immediate or late clinical consequences [1–5, 7, 8].

1.4 Newborns without her mother's serology

Another different possible scenario could be to receive a child, without her mother's serology. In this stage, follow them up, doing an IgG toxo test every three months until they are one-year-old [1, 2]. What information is obtained?:

1. At the beginning, during the first six months, the same mother's serology is observed. The specific IgG crosses placenta so the results obtained will be a copy of their mother's antibody titer. But when the baby's immune system matures, mother's antibodies disappear. This is the moment to be more careful and make sure the parasite does not take advantage of it. This is the reason why the child's IgG antibodies, until she/he is one year old, must be followed (as shown in **Figure 2**) [1, 2].

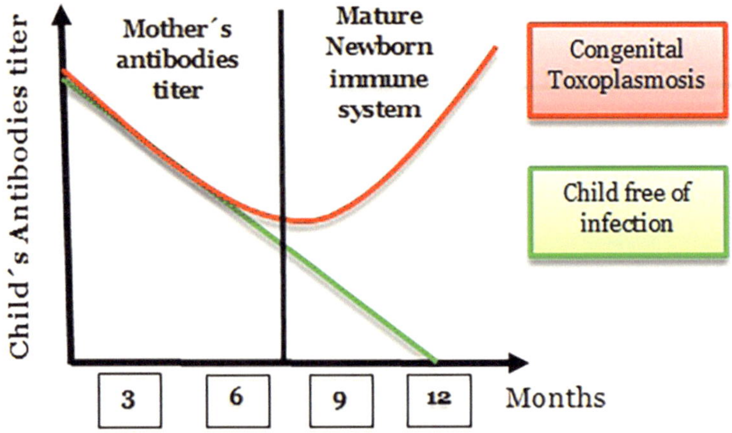

Figure 2.
Newborns without her mother´s serology. Note: Own elaboration.

2. A congenital toxoplasmosis is suspected in the newborn when the IgG titer is significantly higher than the mother's and the IgM is positive. The child is considered free of infection when the IgG is negative at age one (as illustrated in **Figure 2**). This might be the gold standard for laboratories, having peace of mind that a good work was done, and the mission was fulfilled [1, 2].

3. When a prenatal infection is suspected, an hemogram, hepatogram, and a urine analysis also should be applied as the initial control and value systemic compromise of the infection [1, 2].

2. Kinetics of immunologic response

At the beginning of toxoplasmosis disease, during the first-week post-infection, IgM starts to appear. At this time, the role of Sabin Feldman test is fundamental because it detects IgG and IgM together. The next immunoglobulin (Ig) to appear is IgE, then IgA. IgG appears within the first month, stabilizes two or three months after infection has started, and persists throughout life (as shown in **Figure 3**) [1–5, 7, 8, 11].

2.1 IgG detects different Toxoplasma *gondii* antigens. First antigens to be detected by the IgG are membrane antigens and later intracytoplasmics ones [1].

2.2 To detect IgE, IgA and IgM ISAGA's test is needed (see Diagnostics available tests).

2.3 IgM displays a special behavior in this disease. It can be found for more than two years [1, 3, 7, 8, 11, 18]. Hence, if it is found, be careful how its appearance is interpreted. Complement that finding using other tests like specific IgA and IgE must be done to confirm a SC.

2.4 IgA is found during about six months, if it is positive, it is a correct way to confirm a SC [1–5, 8, 11].

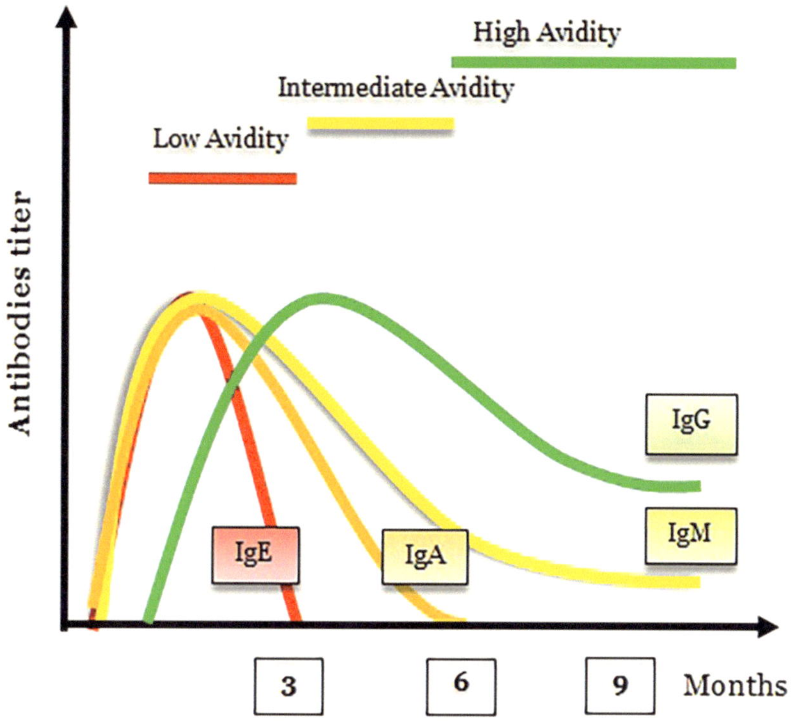

Figure 3.
Kinetics immunologic response. Note: Own elaboration.

2.5 IgE falls fast. Only during the first three month of *Toxoplasma gondii* infection, it can be detected, but whenever is found, it also confirms the SC [1–5, 8, 11].

2.6 IgE and IgA together increase the chances of detecting seroconversion earlier [1–5, 8, 11]. Therefore, it is important not to waste time and confirm those questionable samples as soon as possible.

3. Diagnostic available tests

There are direct and indirect methods to detect a Toxoplasmosis infection:

3.1 Direct methods

They are based on identifying the presence of *Toxoplasma gondii* parasite or their nucleic acid. The techniques within these methods are fresh observation, histological stains, histopathological analysis, PCR, PCR real time, parasite isolation.

The materials available to use are newborn blood, umbilical blood cord, placenta, amniotic fluid, LCR, vitreous humor, biopsies, and other fluids [1, 8].

3.1.1 PCR

It is based on detecting the parasite nucleic acid using the polymerase chain reaction.

3.1.2 Parasite isolation

It consists of detecting the parasite in biological samples (cell culture or mouse inoculation).

3.2 Indirect methods

All of these tests evidence specific antibodies. There are screening tests: indirect hemagglutination (HAI), direct agglutination (DA), immunoassays, and confirmatory tests: Sabin Feldman, ISAGA M, A and E, IFI IgG and IFI IgM, and avidity test.

All of them have their advantages and disadvantages (see **Tables 1** and **2***)* [1, 3, 8, 11]. *The choice depends on the kind of laboratory and resources available.*

3.2.1 Screening tests

This tests are used in Routine Laboratories (**Table 1**)

3.2.1.1 HAI

This technique is not recommendable. However, many laboratories still use it; therefore, it is necessary to warn that it detects late infection. It is useful only after three months of the onset of the infection.

3.2.1.2 AD

Direct agglutination consists of mixing a stabilized suspension of *Toxoplasma gondii* or latex particles sensitized with parasitic antigens, with the patient serum and observing agglutination reaction.

3.2.1.3 Immunoassays

The basis of these tests is to measure the specific union antigen-antibody IgG, by adding a detection complex. The antigen is attached to different supports. That is why

Test	Automatizable	Staff dependant	Need special equipment	Cost	Sensitivity for SC	Specificity
HAI	No	Yes	No	Low	Low*	Low
AD	No	Yes	No	Low	Low*	Low
ELISA	Var	Var	Var	Low	Var	Var
QML, CLIA, ELFA	Yes	No	Yes	Var	High	High
Avidity	Var	Var	Var	Var	High**	High

Var: Variable: Depends on the cost/benefit of the laboratory.
Note: Own elaboration. *Detects late infection.*
**When the result is low and pregnancy is before the 16th week.*

Table 1.
Screening tests. Routine Laboratories [1].

Test	Automatizable	Staff dependant	Need special equipment	Cost	Sensitivity for SC	Specificity
Sabin Feldman*	No	Yes	No	Low	High	High
IFI IgG	No	Yes	Yes	Low	High	High
IFI IgM	No	Yes	Yes	Low	Low	High
ISAGA M	No	Yes	No	Low	Var	High
ISAGA A	No	Yes	No	Low	High	High
ISAGA E	No	Yes	No	Low	High	High
Western Blot	No	Yes	No	High	High**	High
PCR in House, RT PCR	Var	Var	Yes	Var	Var	High
Isolation	No	Yes	No	Low	High	High

Var: Variable: Depends on the laboratories cost/benefits.
*Note: Own elaboration. *Reference reaction.*
***For paired samples mother-child.*

Table 2.
Confirmatory tests. Reference Laboratories [1].

there are a lot of available types and brands of immunoassays (ELISA, QML, CLIA, IFI), their variety depends on the following:

3.2.1.3.1 The kind of antigen that they use

There are membrane antigens that appear first (p30 (SAG-1), p22 (SAG-2)) and cytoplasmic antigens that can be detected later (p35 (GRA-8), ROP-1, GRA-6, GRA-7). According to the antigens that the immunoassay uses, the specific IgG can detect them within one or three weeks after infection has started [1]. Deciding which test to choose is one of the most important steps.

3.2.1.3.2 The support that is being used

Consist on the base where the antigens or antibodies are sustained. They can be microplates (ELISA), microparticles (QML, CLIA), slides (IFI), latex particles (AD), cells (HAI). The smaller they are, the more sensitive the test will be.

3.2.1.3.3 The detection complex is the other ítem to be considered

Conjugate-enzyme (ELISA), fluorescence (IFI), luminiscencia (CLIA), quimioluminiscencia (QML).

The sensitivity and specificity of different assays available depend on the combination of all of these variables; therefore, the choice made will depend on the cost/benefit of the laboratories available resources.

But the most important after selecting the tests is to follow the insert and do the test respecting it, doing exactly what it recommends and always use quality controls, internal and external ones, to compare the results and be sure that all samples have a certain result.

3.2.2 Confirmatory tests

These tests are used in Reference Laboratories (**Table 2**).

3.2.2.1 Sabin Feldman dye test (SF)

The test is based on complement-mediated cytolysis of antibody-coated live Toxoplasma gondii tachyzoites, which is indicated by their inability to take up methylene blue stain [19–22].

This test deserves a special topic and is a reference method for serodiagnostic *Toxoplasma gondii* infection. Its importance is based on being the only one that detects IgG and IgM together. Therefore, it is the most specific and sensitive of all the indirect methods. There are cases that disagree between values comparing SF and other tests, because it is the first one to render a positive result. This test makes the difference to detect the seroconversion as soon as possible.

We are one of the laboratories that still use it as a routine method. A multicenter study of prevention of congenital toxoplasmosis, based on confirmed equivocal or indeterminate samples from other routine laboratories, gets as result that the 46% of the samples that were serodiagnostic as possible seroconversions in routine laboratories, were defined as cronics in our reference laboratory using SF and ISAGA M method. This means that all of those women would have been treated when not necessary [2].

The problems of most laboratories that stop using SF were maintained live tachyzoites of *Toxoplasma gondii*, each country's animal regulations, trained staff, dangerous to laboratory workers, costs, automation, interlaboratory standardization, accessibility of Western blot, better performance of new tests.

Because of them, probably the future of the SF dye test in reference laboratories could be changed for cellular culture, if possible, which is a valid alternative and could be the way to change the SF dye test without losing its advantages and quality.

3.2.2.2 ISAGA

It consists of capturing the specific antibody (IgM, IgA, or IgE) and then adding a stabilized suspension of *Toxoplasma gondii* to evidence specific Igs with direct agglutination.

3.2.2.3 Avidity Test

The avidity test measures the power of the union antigen-antibody. When it is recent, the union is weak, but when it is not recent, the union is strong. We could understand it like a traffic light (illustrated in **Figure 3**): When the result is low, the union is weak and it means high probability of seroconversion within the past three months (red light); if it is intermediate, then we must be alert and study the combination of the other different serologies we have (yellow light); and when the result is high, the union is strong and can be reassured that it is a more than four months old infection (green light).

It is important to consider that this test gives useful information only until the 16th week of pregnancy. Any time after the 16th week, the avidity test loses its importance [1, 3, 4, 8, 11]. *Another consideration is that there are many brands of avidity tests available. They each have their own benchmarks and sometimes the results do not match up. Therefore, it is very important to consider the other confirmatory tests (see Available diagnostic tests) at the time of the final decision.*

4. What should be done in Routine Laboratories for screening?

Here, the big question to answer what must be done in Routine Laboratories for screening, is: Only IgG? IgG and IgM?

Well, the answer is not an only one, the answer depends on the resources and the cost/benefit analysis.

Do only IgG tests should be done as frequently as possible, which means between every month and every three months, mínimum [1–5, 9, 10]. This is the only way to detect an early SC.

If IgM is added, chances of detecting seroconversion are much better [1–5] because those cases where the SC has just begun will be able to be detected.

This is certainly the best way to work. The decision will depend on the prevalence [2, 5, 6], the patient's knowledge of this disease and their access to primary preventive measures (see Preventive Medicine) and laboratory resources.

5. What should be done in Reference Laboratories?

As a continuity on the previous topic, here is the great difference.

While Routine Laboratories can choose a lot of brands of kits, and do only IgG depending on their resources and cost/benefit analysis, Reference Laboratories cannot. That is what makes the difference.

Reference Laboratories must do, not only IgG and IgM, but also the most sensitive ones and have the best brands. Besides that, must do ISAGA´s Tests ISAGA M, ISAGA A, and ISAGA E, Western blot, PCR [8, 23], and if possible Sabin Feldman or Cellular Culture and sequence [12–17, 19–23].

This difference is based on accurately diagnosing a seroconversion. Cannot report a doubt. If there is a doubt, the best they can do is to recommend retesting after two or three weeks, to allow the immunological system to mature. This is when a SC has just started and the titer of IgM is not enough to confirm it. Meanwhile, they should suggest doctors in charge start medicating them, as a preventive measure, until the result of the next sample is obtained.

6. Preventive medicine

Preventive medicine is another very important task to detect SC earlier than anything. There are three ways to do this:

6.1 Primary Preventive Medicine

There are many things to do to prevent *Toxoplasma gondii* infection. Precautionary measures consist of [1, 3, 4, 7, 8]:

1. Washing hands before handling foods.

2. After handling raw meat, wash hands with water and soap.

3. Not eating undercooked meat. It must be cooked at more than 66°C.

4. Not eating sausages.

5. When handling foods, a different cutting board for meat and raw vegetables must be used.

6. Cleaning the surface and utensils after using them with raw foods.

7. Cleaning fruits and vegetables very carefully. They must not be eaten if it is not known how they were washed.

8. Avoiding contact with unknown cats.

9. Wearing gloves for cleaning domestic cat excrement and washing hands carefully after.

10. Wearing gloves and washing hands carefully after practicing gardening.

There are a lot of things laboratories can do to make pregnant women aware of these precautionary measures and help them prevent *Toxoplasma gondii* infection. There are many ways to do this and it will depend on empathy, enthusiasm, and imagination. Some ideas are as follows:

- Add them in your report.

- Make brochures and give them to pregnant women at their gynecological consultation.

- Use all the social media networks (Internet, Twitter, Instagram, Web pages).

- Use your internal network at your job.

- Publish Posters, etc.

Congenital toxoplasmosis can cause miscarriage or stillbirth. In many cases, infants appear healthy at birth but they may develop adverse sequelae of the infection later in life, including decreased vision or blindness, decreased hearing or deafness, and mental and psychomotor retardation.

The idea is to focus on these measures and to make pregnant women aware of the Toxoplasma gondii infection and of the consequences they could face if this infection is not taken seriously.

As an example, we work in a Reference Laboratory Country and we receive samples from many cities in the country, to help them confirm those difficult results, with values near the cut off that routine laboratories cannot resolve because of their limited techniques. We are working on a prevention toxoplasmosis program named "CONTENER," it means "Containing," to expand our services to help more laboratories and concerned patients' get an accurate result. When this happens, our mission to help getting a final and accurate diagnosis to apply the right treatment as soon as possible is accomplished. "CONTENER" is also an Spanish acronym for "Contain against Toxoplasmosis in newborns and pregnant women".

6.2 Secondary preventive medicine

If primary prevention did not work and a SC has happened, this preventive option appears. The secondary prevention is directed to protect the embryo or fetus when the mother has been infected during pregnancy. It is based on antiparasitic treatment until delivery. The most important thing in this situation is to start it as soon as possible. The sooner it starts, the more effective it will be [1, 3, 8].

The next step is to do serologic analysis on the paired mother-child. The ideal test to use is Western blot, but any IgG- and IgM-specific paired serology achieves. Some Reference centers suggest detection of DNA parasites using the blood cord of the newborn as an important complementary fact of great specificity and we also recommend isolation of the parasite in cord blood and placenta if it is possible [12–17].

6.3 Third preventive medicine

This preventive measure is used when the other two fail. This next step is to make a referral to a specialist doctor who understands this disease and its importance to be medicated as soon as possible to prevent the immediate and late clinical consequences and perform all of the necessary studies and controls.

7. Laboratories role in pregnancy and newborn

After reviewing everything laboratories can do about these topics, it is necessary to highlight the following:

1. Study pregnant women serology before and during pregnancy with frequent controls [1, 2, 9, 10].

2. Use the best tests available with the resources at hand.

3. Do IgG and IgM if IgG is positive [1–4, 7, 8, 11].

4. Send samples or refer them to specialized laboratories when cannot obtain an accurate result whenever a positive IgM result is obtain.

5. Have a fluid communication with the doctor in charge following pathologic serology results.

6. Be in contact with the pediatrician in charge of the newborn.

7. Do Preventive Medicine [1, 3, 4, 7, 8].

8. Spread as many precautionary measures as possible

9. Check periodic and paired serology evolution of those kids born from *Toxoplasma gondii* infected women, until they are one-year-old [1, 2, 4].

8. Laboratories role in other clinicals group.

8.1 Immunocompetent patients

Most toxoplasmosis infections are asymptomatic or have mild nonspecific symptoms, so these patients who have a healthy immune system, probably do not need to be diagnosed. This is the reason why occasional findings of circulating specific antibodies are common. Chronic infections are more frequent than acute ones [24] and patients should not require treatment because it is a self-limited disease.

Mild symptoms could be: high fever, astenia, headache, muscle pain, anorexia, cough, throat pain, lymphadenopathies, and splenomegaly (when there is involvement of the mesenteric nodes) and they must be differentiated with other diseases.

The most characteristic symptom that needs to be differentiated is lymphadenitis: In this case, the role of laboratories gains importance in giving an accurate diagnosis to specialist doctors, not to start treatment, but to rule out other pathologies that do need it.

8.2 Immunosuppressed patients

These patients are susceptible to develop diseases that may cause severe sequelae depending on the type of infection. Referring to toxoplasmosis, two types of illness can be developed: severe primary infection and reactivation of a latent infection.

For example, immune deficiency conditions such as AIDS and organ transplantation can cause fatal toxoplasmic encephalitis [24–27]. Hence, the role of laboratories in these cases is fundamental to accompany other diagnostics methods such as RMN and TC.

To diagnose, direct methods in samples, such as LCR, lymph nodes, vitreous humor, bone marrow (see Diagnostic available tests), should be used. Even though these tests do not have high sensitivity, they have a high-positive predictive value (VPP) when the result is positive, and when it is negative (negative predictive value (VPN)) it will alert doctors to a risk of a primary infection. Hence, periodic serological controls and preventive measures should be carried out.

8.2.1 VIH patients

Encephalitis with brain injuries is the most common sequelae when patients are HIV positive and have the disease [24–27]. The role of laboratories is to complement results obtained, with others such as TC and RMN and biopsy.

Direct methods like PCR or isolation can be useful in fluids such as LCR, BAL, biopsies.

Serology results are important to predict risk of infection or reactivation: SF test has a VPN del 99,7% and a VPP of 88% and other serologies, IgG above 1/256 with CD4 lymphocytes less than 150 cel/mm3, ISAGA M 12%, ISAGA A 38% e ISAGA E 25%.

8.2.2 Transplanted patients

Myocarditis, neumonitis, and SNC compromised could happen in transplanted patients with reactivation of toxoplasmosis. Laboratories importance is based on knowing serologies, donor, and recipient, so that doctors in charge can be alert to any risk of infection or reactivation [24].

8.2.3 Ocular toxoplasma infection

Toxoplasmosis is a major cause of retinochoroiditis, especially in individuals with an impaired immune system. It is so important to diagnose a toxoplasma infection when ocular infection occurs, because this acute infection must be treated. The consequences of not doing so, could bring serious complications: When a cyst ruptures, retinochoroiditis presents an intense inflammatory reaction that tends to scarring. In immunodeficiency conditions, necrosis occurs and a retinal detachment, uveitis, and then secondary glaucoma could be presented [24, 26].

The laboratories roles are based on direct detection of the parasite in vitreous humor and confirmatory serology (see Diagnostic available tests). The high VPP of these results with ocular lesions helps treatment as soon as possible to avoid the worse lesions.

8.2.3.1 Hodgkin disease and other lymphomas

Must be also differentiated from toxoplasmosis. Laboratories direct methods in any material like biopsia of nodes is as important as indirect methods, both to rule out or to confirm toxoplasma infection.

9. Conclusions

In conclusion, the laboratories mission should be:

9.1 In pregnancy and newborns

1. Prevent parasite *Toxoplasma gondii* from infecting pregnant women. If a parasite passes the placenta without treatment, the risk of infection is 40-50% [2].

2. Do the best as possible with the resources at hand, to monitor seronegative pregnant women to make sure they do not become infected. Keep in mind that the most common maternal infections are asymptomatic.

3. Have a fluid communication with the attending physician to alert them once a pathological result is found.

4. If a pregnant woman gets *Toxoplasma gondii* infection, prevent the parasite from passing the placenta and infecting the future newborn by diagnosing accurately, in order to administer the appropriate medicine. Pregnant women and fetuses must receive treatment to minimize clinical consequences [1–4, 7, 8].

5. Take care of newborns until they are one-year-old and their immune system matures. The majority of these babies are born without symptoms [1–4].

6. If they do not receive treatment, the possibilities of having immediate or late clinical consequences could be up to 80%, which would make those families' future very difficult [1, 2, 8].

9.2 In other clinical groups

The role of laboratory diagnosis through direct methods on any material received and indirect methods (screening or confirmatory serologic assays) plays a main role in the differential diagnosis from other diseases. This is where the major effort must be made:

9.2.1 Immunocompetent patients

Laboratories become important to rule out *Toxoplasma gondii* infection so doctors can be able to focus on other pathologies.

9.2.2 HIV

Laboratories must follow HIV patients for the risk of infection or reactivation, when the count of linfocitos CD4 is less than 150/mm3. Therefore, the serology results are a complement when an increase of the titer of IgG and IgM appears. If treatment responds favorably in seroreactive patients, it also reinforces diagnosis [24–27].

9.2.3 Transplant

Bone Marrow or solid organ transplantation must be periodically monitored by laboratories; in order to be alert for reactivation and when donor and/or receptor are serology positive previous to the transplant.

9.2.4 Chorioretinitis

Laboratories should play an important role in doing PCR and/or isolation in vitreous humor, besides confirmatory serology such as IgG, ISAGA M, ISAGA A, and ISAGA E. It is very important to notify doctors to start treatment as soon as possible and thus avoid complications [24].

9.2.5 Hodking disease and other lymphoma

Confirming *Toxoplasma gondii* infection or ruling it out should be the purpose of laboratories. Doctors must be aware of the diagnosis to know what steps to follow.

9.3 Choice of tests

1. No matter what combination of tests is used, the best must be done with the resources at hand.

2. Not every laboratory has the same budget to afford the best brand. It is of utmost importance to responsibly follow the test's insert instructions, and to use internal and external quality controls.

3. Refer to reference laboratories; those samples with gray area or unclear results for obtaining an accurate and final result.

If laboratories do this, they can be rest assured that they did their best in making sure future children are born free of clinical consequences or ruling out the disease or confirming toxoplasmosis from any other pathology.

9.4 Result information

1. If there are any doubts about the result and there is no reference laboratory within reach, the report should include information such as: "gray zone" and under observations, suggesting repeating the test in two or three weeks . When the new sample arrives, comparing both sample titles.

2. Use observations to communicate to doctors any advice: for example, "Retesting recommended", "Expand with more specific tests," or "The result corresponds to the cut-off value."

To conclude: All of these objectives are equally important: From getting involved in primary prevention to doing the best job possible.

In other words, working together as a team with the attending doctors, focusing on responsibility, efficiency, quality, and commitment in order to provide the best service to patients; are certainly key to "The important role of laboratories" "Towards new perspectives on Toxoplasma gondii."

Acknowledgements

I would like to thank:

My dear childhood friend Patricia Negritto, her dedication and commitment just for empathy, helping and accompanying, motivated, and moved me.

D. Sc Liliana Carral and D. Sc. Federico Kaufer for opening the doors to the world of toxoplasmosis for me and for their support, trust, and teaching.

Dr. Cristina Freuler for her support and confidence.

D.Sc. Jorgelina Blejer: for her empathy, motivation, teaching, and professionalism.

Cecilia Miñones for her help, friendship, and professionalism.

Silvia Sabo for her predisposition, commitment, help, and understanding.

My coworkers Matías Messina and Ricardo Medina for making me enjoy every day of my work. It is very easy and pleasant to work with them.

Guillermo Omar Di Stasio for inspiring me with his book "Tu ventaja diferencial" and helping me in a special moment.

Tomy: for being the best son, so empathetic, kind, funny, charming, and companion.

My father, Jorge Schneider, for teaching me above all, that there is no obstacle that cannot be overcome.

My mom who accompanies me from wherever she is.

And to all my friends who support me, motivate me, and teach me every day how important it is to have people in life with whom to share moments, good, and not so good ones (Zule, Vero, Lily, Gaby, Faby, Mary, Marce, Tere, Nancy, Lore, Normi, Bea).

Thank God for always be by my side and showing me the signs where I should go.

Conflict of interest

The author declares no conflict of interest

Author details

María Vanessa Schneider
Hospital Alemán, CABA, Argentina

*Address all correspondence to: vaneschneider2021@gmail.com

IntechOpen

© 2022 The Author(s). Licensee IntechOpen. This chapter is distributed under the terms of the Creative Commons Attribution License (http://creativecommons.org/licenses/by/3.0), which permits unrestricted use, distribution, and reproduction in any medium, provided the original work is properly cited.

References

[1] Durlach R, Freuler C, et al. Consenso argentino de toxoplasmosis congénita 2020. Medicina (B Aires). 2021;**81**:257-268

[2] Carral LA, Kaufer F, Durlach R, et al. Estudio Multicéntrico para la Prevención de la Toxoplasmosis Perinatal en Buenos Aires. Medicina (B Aires). 2008;**68**:417-422

[3] Durlach R, Kaufer F, Carral L, et al. Consenso argentino de toxoplasmosis congénita. Medicina (B Aires). 2008;**68**:75-87

[4] Carral L, Kaufer F, Olejnik P, Freuler C, Durlach R. Prevención de toxoplasmosis congénita en un hospital de Buenos Aires. Medicina (B Aires). 2013;**73**:238-242

[5] Messina M, Schneider MV, Carral L. Prevalencia de anticuerpos anti-Toxoplasma *gondii* en embarazadas en siete Hospitales de Buenos Aires. Comparación entre 2006 y 2017. Medicina (B Aires). 2019;**79**:433-437

[6] Kaufer F, Carral L, Messina M, et al. Prevalencia de anticuerpos anti-Toxoplasma gondii en hemodonantes en la Ciudad de Buenos Aires, desde 1967 a 2017. Medicina (B Aires). 2017;**77**:475-480

[7] L. Paris. Toxoplasmosis. EMC-Tratado de Medicina. 2009;**13**(4):1-7

[8] Baquero-Artigao F, del Castillo MF, Fuentes Corripio I, et al. The Spanish Society of Pediatric Infectious Diseases Guidelines for the diagnosis and treatment of congenital toxoplasmosis. An. Pediatr. (Barc). 2013;**79**(2):116.e16

[9] Aspöck H, Pollak A. Prevention of prenatal toxoplasmosis by serological screening of pregnant women in Austria. Scandinavian Journal of Infectious Diseases. 1992;**84**:32-37

[10] Thulliez P. Screening programme for congenital toxoplasmosis in France. Scandinavian Journal of Infectious Diseases. 1992;**84**:43-45

[11] Contreras, M del C. et al. Diagnostic utility of ELISA IgG, IgM, IgA and ELISA IgG avidity in recent and chronic toxoplasmosis. Boletin Chileno de Parasitologia. 2000;**55**(1-2):17-24.

[12] Pardini L, Carral LA, Bernstein M, et al. First isolation and molecular characterization of Toxoplasma gondii from a human placenta in Argentina. International Journal for Parasitology. 2014;**63**:470-472

[13] Carral L, Kaufer F, Pardini L, Durlach R, Moré G, Venturini MC, et al. Toxoplasmosis congénita: Serología, PCR, aislamiento parasitario y caracterización molecular de *Toxoplasma gondii*. Rev Chilena de Infectología. 2018;**35**:36-40

[14] Bernstein M, Pardini L, Moré G, Unzaga JM, Su C, Venturini MC. Population structure of *Toxoplasma gondii* in Argentina. Infection, Genetics and Evolution. 2018;**65**:72-79

[15] Pardini L, Bernstein M, Carral LA, et al. Congenital human toxoplasmosis caused by non-clonal Toxoplasma gondii genotypes in Argentina. International Journal for Parasitology. 2019;**68**:48-52

[16] Bernstein M, Pardini L, Carral L, et al. Genotipos de Toxoplasma *gondii* en casos de Toxoplasmosis congénita en Argentina. In: XVI Congreso de la

Sociedad Argentina de Infectología-Mar del Plata. 2017

[17] Pardini L, Carral L, Bernstein M, et al. Nuevo aislamiento atípico de Toxoplasma *gondii* en un caso de toxoplasmosis congénita aguda. In: XVI Congreso de la Sociedad Argentina de Infectología-Mar del Plata. 2017

[18] Gras L, Gilbert RE, Wallon M, Peyron F, Cortina-Borja M. Duration of the IgM response in women acquiring Toxoplasma gondii during pregnancy: Implications for clinical practice and cross-sectional incidence studies. Epidemiology and Infection. 2004;**132**:541-548

[19] Sabin AB, Feldman HA. Dyes as microchemical indicators of a new immunity phenomenon affecting a protozoan parasite (Toxoplasma). Science. 1949;**108**:660-3. 88

[20] Walton BC, Benchoff BM, Brooks WH. Comparison of the indirect fluorescent antibody test and methylene blue dye test for detection of antibodies to Toxoplasma gondii. American Journal of Tropical Medicine and Hygiene. 1966;**15**:149-152

[21] Reiter-Owona I, Petersen E, Joynson D, Aspöck H, Dardé ML. et al. The past and present role of the Sabin-Feldman dye test in the serodiagnosis of toxoplasmosis. Bulletin of the World Health Organization. 1999;77(11): 929-935. https://apps.who.int/iris/handle/10665/267936

[22] Acharya Tankeshwarin. Sabin-Feldman dye test: Principle and Limitations. In: Parasitology https://microbeonline.com/sabin-feldman-dye-test/. [Last Updated October 4, 2022]

[23] Belaz S, Gangneux JP, Dupretz P, Guiguen C, RobertGangneux F. A 10-year retrospective comparison of two target sequences, REP-529 and B1, for Toxoplasma gondii detection by quantitative PCR. Journal of Clinical Microbiology. 2015;**53**:1294-1300

[24] Botero D, Restrepo M. Parasitosis humanas. CIB (Medellin, Colombia). 1994;**9**:235-245

[25] Camila Pantoja-Ruiz. Toxoplasmosis in the central nervous system: Review of the pathology, diagnostic approach and treatment. Acta Neurol Colomb. 2021;37(1):Supl. 1:141-147

[26] Saadatnia G. A rewiew on human toxoplasmosis. Scandinavian Journal of Infectious Diseases. 2012;**44**(11):805-814

[27] Vidal JE. HIV-related cerebral toxoplasmosis revisited: Current Concepts and controversies of an old disease. Journal of the International Association of Providers of AIDS Care (JIAPAC). Review August 2019;**18**:1-20. DOI: 10.1177/2325958219867315